III International Scientific and Practical Conference "Advanced Building Materials and Technologies" (MCMT2020)

IOP Conference Series: Materials Science and Engineering
Volume 911

Kaliningrad, Russia
26 - 29 May 2020

ISBN: 978-1-7138-3166-2
ISSN: 1757-8981

Printed from e-media with permission by:

Curran Associates, Inc.
57 Morehouse Lane
Red Hook, NY 12571

Some format issues inherent in the e-media version may also appear in this print version.

Institute of Physics
Dirac House, Temple Back
Bristol BS1 6BE UK

Phone: 44 1 17 929 7481
Fax: 44 1 17 920 0979

techtracking@iop.org

Additional copies of this publication are available from:

Curran Associates, Inc.
57 Morehouse Lane
Red Hook, NY 12571 USA
Phone: 845-758-0400
Fax: 845-758-2633
Email: curran@proceedings.com
Web: www.proceedings.com

TABLE OF CONTENTS

ACTUAL ISSUES OF BUILDING MATERIALS SCIENCE

INTENSIFICATION OF THE PROCESSES OF PREPARATION OF DRILLING AND CEMENT MORTARS USING VIBRATION JET ACTIVATION METHODS 1
S V Rikkonen, M A Dmitrieva, V N Leitsin, C V Ponomarev, A V Azin

PORE COLMATATION IN CASE OF LIQUID CORROSION OF CONCRETE 7
S V Fedosov, V Eu Rumyantseva, V S Konovalova, A S Evsyakov

CEMENT BASED FOAM CONCRETE WITH HARDENING ACCELERATORS 13
A B Steshenko, A I Kudyakov, N E Ryabtseva

APPROACH TO EVALUATING THE CHANGE OF PROPERTIES OF THE GEOSYNTHETIC MATERIAL USED TO STABILIZE THE MARINE LANDSCAPE SLOPES 19
V N Leitsin, A O Tovpinets, B V Chubarenko, D A Domnin, E E Esiukova, E M Burnashov

ANALYSIS OF EARLY STRUCTURE FORMATION OF ANHYDRITE BINDERS FROM SECONDARY RAW MATERIALS USING MICROCALORIMETRY 25
L A Anikanova, O V Volkova, A I Kudyakov, V A Lotov, D M Pervushina

EXPERIMENTAL STUDIES OF THE EVOLUTION OF STRUCTURAL AND MECHANICAL CHARACTERISTICS OF CEMENT STONE IN THE PROCESS OF HYDRATION 30
M A Dmitrieva, A V Sharanova, V N Leitsin

OPTICAL PROPERTIES OF SOL-GEL COATINGS CONTAINING SILVER NANOPARTICLES 35
S N Stepin, T A Xolmurodov, O O Mirzaev, V E Katnov

ACTUAL ISSUES OF THE DESIGN OF BUILDING STRUCTURES

INSITU DETERMINATION OF CONTINUOUS EVOLUTION IN SPACE AND DURING THE STRAIN FIELDS ON STEEL PLATE WITH THE STRESS CONCENTRATOR 41
V A Vlasov, A S Plyaskin, A A Klopotov, A I Potekayev, A M Ustinov, Yu A Abzayev, A Kurgan, V E Bunkov, N A Matskevich

EVALUATION OF STRAIN-STRESS STATE OF VERTICAL TANK REINFORCED BY CARBON TYRE BASED ON NUMERICAL RESEARCHES IN ANSYS PC 47
V A Vlasov, A A Klopotov, A S Plyaskin, V E Bunkov, A M Ustinov, M E Savintceva

ROLE OF TEMPERATURE FACTOR IN THE WORK OF A CELLULAR REINFORCED CONCRETE ICE-RESISTANT PLATFORM 53
V A Zimnyukov, M I Zborovskaya

ALGORITHM FOR DESIGNING BUILDING CONSTRUCTIONS EXPRESSED BY NONLINEAR FUNCTIONS 61
S N Volkova, A V Shleenko, E E Sivak

ASSESSMENT OF ENERGY EFFICIENCY OF YIELDING SUPPORTS FOR REINFORCED CONCRETE BEAMS UNDER DYNAMIC LOADING 67
O G Kumpyak, Z R Galyautdinov, D R Galyautdinov

SIMULATION OF THE STRAIN DIAGRAM OF A SLAG-CONCRETE ELEMENT SUBJECT TO BENDING 72
N Chernousov, B Bondarev, V Sturova

VACUUM FILTRATION ON A THIN VERTICAL FILTERING BAFFLE 80
E Yu Kurochkin, N Yu Kurochkina

ECOLOGICAL AND ECONOMIC ASPECTS OF CONSTRUCTION IN FORMATION OF A COMFORTABLE URBAN ENVIRONMENT

DESIGNING NEW SUSTAINABLE ANTI-LANDSLIDE SYSTEMS UNDER CONSTRAINED CITY CONDITIONS 85
T P Kasharina, D V Kasharin, D A Polyakov

QUALITY ECOLOGICAL ASSESSMENT OF THE URBANIZED TERRITORIES' ENVIRONMENT 90
L A Kaverzina, V A Nikiforova, V V Dorofeeva

FORMATION OF A COMFORTABLE URBAN ENVIRONMENT, TAKING INTO ACCOUNT THE RECONSTRUCTION, RESTORATION OF BUILDINGS AND STRUCTURES (ON THE EXAMPLE OF THE KALININGRAD REGION) 96
V I Pustovgarov

RISK ASSESSMENT OF INVESTMENT PROJECTS IMPLEMENTED ON THE BASIS OF PROJECT FINANCING 102
I V Bryantseva, N V Voronina

FACTORS OF ENVIRONMENTALLY ORIENTED DEVELOPMENT OF MANAGEMENT TECHNOLOGIES IN CONSTRUCTION 108
I Nuzhina, L Shershova

THE PROCESS AND PROJECT APPROACH AS THE BASIS FOR HARMONIZATION OF DEVELOPMENT MANAGEMENT SYSTEM OF A CONSTRUCTION COMPANY 114
V V Dorofeeva, L A Kaverzina, I P Nuzhina

ANALYSIS OF THE REQUIREMENTS OF NORMATIVE AND TECHNICAL DOCUMENTATION FOR PILING EQUIPMENT 120
S Plekhanova, N Vinogradova

Author Index

IOP Conf. Series: Materials Science and Engineering **911** (2020) 011001 doi:10.1088/1757-899X/911/1/011001

Immanuel Kant Baltic Federal University

MCMT2020

ADVANCED BUILDING MATERIALS AND TECHNOLOGIES

May 26-29, 2020, Kaliningrad, Russian Federation

The III International Scientific and Practical Conference "Advanced Building Materials and Technologies" (MCMT2020) is devoted to considering the latest experience in scientific and innovative activities in the field of building materials and technologies, discussing modern scientific achievements and world trends in the use of local raw materials for the construction industry, the formation of a comfortable urban environment and economic aspects in construction. The conference was organized by the Immanuel Kant Baltic Federal University and became a traditional platform for discussing modern issues of the construction industry development with the invitation of leading experts from Russia, Australia, Armenia, Belarus, Germany, China, Lithuania, Mongolia, Poland and the Czech Republic.

The topics of reports at the conference are presented in three main areas: Actual issues of building materials science, Actual issues of design of building structures, Environmental and economic aspects of construction in the formation of a comfortable urban environment. The following topics are considered: aspects of the development of building materials science, issues of mechanical behavior of elements of building structures and ways of their optimization, important aspects of ecology and safety of development of urban areas, promising approaches to making management decisions and minimizing risks in the construction industry.

The Conference proceedings contain selected articles that represent the significant results of the conference. We are grateful to all authors who recognized MCMT2020 as a suitable forum to present their latest developments and learn about new trends in the field of engineering and construction technology.

In a difficult epidemiological situation, it was possible to unite the world's leading specialists in the construction industry. The organizing committee of the conference expresses its sincere gratitude for all-round assistance and support to the Rector's office of the Immanuel Kant Baltic Federal University, to Rector of the Tomsk State University of Architecture and Building, to the Ministry of Construction and Housing and Communal Services of the Kaliningrad Region, LLC Proftechnologies, LLC Technoservice, Stachema Polska, Stieblich Hallenbau GmbH Germany.

We would like to express our gratitude to all reviewers and topic editors for their important work, which has ensured the high quality of scientific reports.

Last but not least, we sincerely thank the entire team that participated in organizing the MCMT2020 conference.

The chairman of Program Committee
Prof. Maria A. Dmitrieva

Conference Coordinators

Associate Prof. Lidya V. Shershova - vice-chairman of the Organizing Committee, Head of the Center for Precision Construction Equipment, IKBFU, Kaliningrad, Russian Federation.

Prof. Maria A. Dmitrieva - vice-chairman of the Organizing Committee, Head of the Department of Urban Planning, Land Management and Design IKBFU, Kaliningrad, Russian Federation.

Anastasiia V. Sharanova - secretary of the organizing committee, Head of the Laboratory of Building Materials, IKBFU, Kaliningrad, Russian Fedearation.

Peer review declaration

All papers published in this volume of *IOP Conference Series: Materials Science and Engineering* have been peer reviewed through processes administered by the Editors. Reviews were conducted by expert referees to the professional and scientific standards expected of a proceedings journal published by IOP Publishing.

- **Type of peer review:** Open
- **Conference submission management system:** -
- **Number of submissions received:** 21
- **Number of submissions sent for review:** 21
- **Number of submissions accepted:** 21
- **Acceptance Rate (Number of Submissions Accepted / Number of Submissions Received X 100):** 100
- **Average number of reviews per paper:** 2
- **Total number of reviewers involved:** 4
- **Any additional info on review process:** -
- **Contact person for queries:** Sharanova Anastasiia a.sharanova@gmail.com

Intensification of the processes of preparation of drilling and cement mortars using vibration jet activation methods

S V Rikkonen[1], M A Dmitrieva[2], V N Leitsin[2], C V Ponomarev[1] and A V Azin[1]

[1]Research Institute of Applied Mathematics and Mechanics, National Research Tomsk State University, 36 Lenin Ave., Tomsk, 634050, Russian Federation
[2]Immanuel Kant Baltic Federal University, 14 Nevskiy St, Kaliningrad, 236016, Russian Federation

E-mail: dmitrieva_m@inbox.ru

Abstract. The article presents the results of laboratory and field tests of vibro-jet magnetic activation (VJMA) technology in the technological processes of preparation of drilling and cement mortars. Studies have shown a significant improvement in the quality of the mortars, reducing the preparation time and energy intensity of the process. The use of VJMA technology for processing grouting mortar allows increasing the fluidity of the solution and increasing the strength of the cement stone. This facilitates the filling of cement mortar.

1. Introduction

Technological mortars must be prepared with certain properties and in large volumes. Each mortar has its own preparing technology, its recipes and its equipment. All of these preparing technologies are energy intensive. Chemical methods for obtaining the required parameters of process fluids are widely used to reduce production costs; however, this exacerbates the requirement for ecotoxicity. There are physical methods of influencing the initial components of process fluids, which can increase productivity and reduce the ecotoxicity of production. The "dry" activation of concrete components can serve as an example of such a physical method [1].

For the preparation and activation of process fluids, effects of various physical nature are used: mechanical, hydraulic, acoustic, thermal, magnetic, electrical. In this case, the main criterion for the degree of preparation can be considered the amount of energy spent on the process [2–7].

Vibration methods for the intensification of technological processes are used steadily in technology, providing a significant increase in productivity and product quality.

2. Research methods

In works [2-4, 8], an effective technology of vibro-jet magnetic activation (VJMA) of technological and raw fluids was developed, which include oil, cement mortars and drilling mud, polymer compositions, saline solutions, oil sludges and bottom sediments, as well as water-oil emulsions and hydrocarbon fuel. The aim of the proposed technology and devices for its implementation is the practical implementation of the principle of complex integration of a number of well-known physical factors of fluid activation in one device (acoustic field, shear flow of turbulent jets, magnetic processing of the medium), which has no analogues and is confirmed by patents and applications [2, 3].

The paper presents the results of laboratory, experimental and operational tests of devices and technology of VJMA.

2.1. Drilling mud preparation
Flushing fluids in drilling processes can have various physical and mechanical properties that have a significant impact on the entire drilling process and on the condition of the well. The quality of the flushing fluid should provide the maximum mechanical speed of drilling, high quality and minimum cost of drilling in specific conditions [9].

Comparative experiments were conducted in laboratory conditions to determine the feasibility of using VJMA technology of fluid to accelerate the preparation of drilling mud using clay from Dolmatovsky quarry (Tomsk).

Mixing the solution was carried out on a laboratory mixer in an open vessel with a rotation speed of 200 rpm. The radius of the mixer blade is 50 mm. All experiments were performed using distilled water. Assessment of the effectiveness of dispersion of quarry clay was carried out by sedimentation analysis. The density of the solution is 1060 kg/m^3. According to sedimentation nomograms, the drilling fluid preparation time was reduced by 5-6 times with the use of VJMA technology (Figure 1) [2].

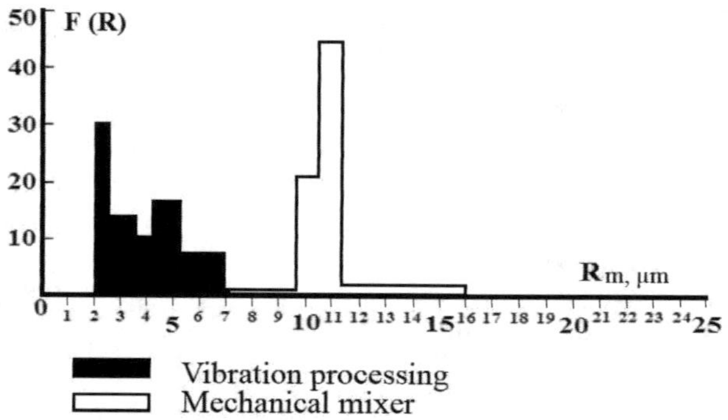

Figure 1. Distribution of particle sizes of clay when using a mechanical mixer and a laboratory vibrator VL-1

Field tests were conducted at well No. 69 BK on the offshore stationary platform OSP-3 of the "White Tiger" field of the "Vietsovpetro" joint venture (Socialist Republic of Vietnam) [10].

Four vibrators of electromagnetic activation VEMA were located in the reservoir for the preparation of the drilling mud (Figure 2). The tests were carried out after the descent of the 245 mm casing in the process of transferring the drilling mud with a density of 1.16-1.19 g/cm^3 to the drilling mud with a density of 1.7-1.73 g/cm^3. The total electric drive power of the mixers is 15 kW. The total electric power of VEMA is 1.2 kW. The control preparation of the drilling mud was carried out in a separate reservoir, where two traditional mechanical mixers were installed.

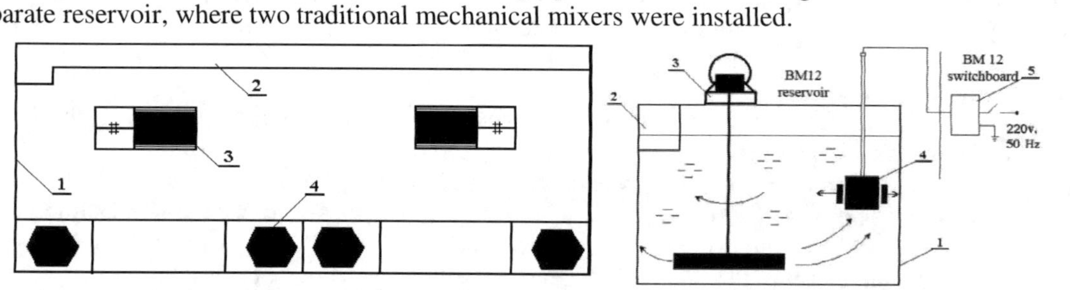

1 - reservoir; 2 - gutter; 3 - mechanical mixers with electric drive; 4 - VEMA-0.3; 5 - junction boxes

Figure 2. Scheme of the vibrators installation in the reservoir

An analysis of the test results shows that all parameters of the drilling mud prepared with VJMA technology are better than the parameters of the drilling fluid prepared in the traditional way. The density in the reservoirs is the same, but in the reservoir with vibrators the basic viscosity is 1.8 times lower. This is explained by the fact that due to the high shear rates and the acoustic field created by the VEMA, a better dispersion of the drilling fluid occurred. Water loss decreased by 11%.

Structural and mechanical properties of the drilling mud are characterized by the thixotropy coefficient Kt = SShS10/SShS1, where SShS1 is the static shear stress at rest, SShS10 is the static shear stress after 10 minutes. According to the recommendation given in [10], the value of Kt should not exceed 2. When using the technology of vibrojet magnetic activation, the value of Kt is 2.

To prevent possible complications when drilling wells in clay and clay-containing rocks, it is proposed to limit the maximum value of the pH of the flushing fluid to 9.2 [10]. When using VJMA, pH = 9.0.

The mud preparation time in the control reservoir was 30% longer than in the reservoir with vibrators. The reduction in drilling mud preparation time using 4 VEMA vibrators was 1.8 hours in relation to 6 hours in the control reservoir. The increase in the number of VEMA will allow reducing the preparation time to a greater extent and increasing the dispersion of the drilling fluid. Comparative parameters of the drilling mud before the experiment and after the experiment with and without vibrators are presented in Table 1.

Table 1. Drilling mud parameters.

Drilling mud parameters	Before testing	After testing	
		Control reservoir	Reservoir with vibrators
Density (g/cm^3)	1.19	1.73	1.72
Viscosity (sec)	50	61	34
Water loss (cm^3)	4.5	4.5	4.0
Crust (mm)	1.0	1.0	1.0
SShS, 1/10	20/50	35/71	20/40
Kt = SShS10/SShS1	2.5	2.028	2
pH	10.5	9.5	9.0

According to the results of field tests, the dependence of reducing the time for preparing the drilling mud T= f(N) is obtained. Using this dependence, it is possible to estimate the required amount of VEMA to reduce the time required to prepare the drilling mud (Figure 3). Figure 3 shows the dependence of the decrease in energy intensity of the drilling mud preparation process ΔW = f(N). The energy intensity of the process of preparation (weighting) of the drilling mud using four VEMA decreased by ΔW = 22 kWh.

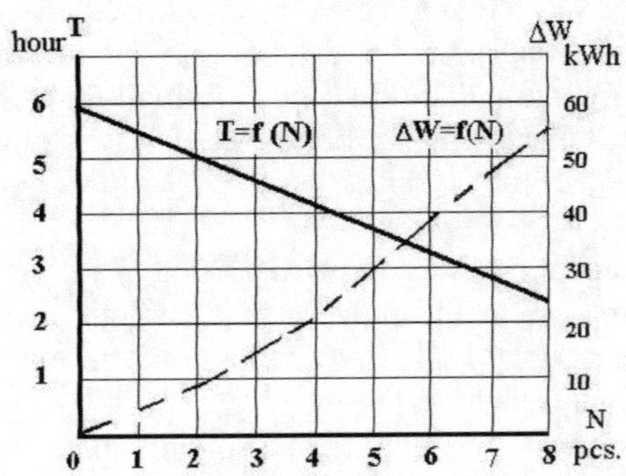

Figure 3. Experimental dependences of reducing the preparation time and reducing energy intensity on the number of vibrators in the process of preparing the drilling mud.

2.2. *Vibration processing and preparation of cement mortar*

One of the most complex processes for fixing wells is their cementing, which includes: preparing a grouting mortar, pumping it into a pipe string, and forcing it into the annulus of the well.

Cementing of oil and gas wells is the final stage of the preparation of the oil rig for operation. The scope of work is aimed at ensuring the maximum service life of the structure. This is due to the following reasons: a) the need to isolate each oil and gas region; b) the need to exclude the possibility of mixing raw materials and water from different layers; c) compliance with the requirements governing the protection of a metal pipe surface [1, 12, 13].

One of the ways to improve the quality of casing cementing during well drilling is to maintain the specified technological parameters of the grouting mortar during its preparation and pumping into the annulus. The use of VEMA in the technological chain will eliminate the drawbacks of the preparation of the mortar, not only by eliminating stagnant zones in the averaging tank, but also by improving the properties of the cement mortar (increasing the spreadability of the mortar, hardening the cement stone).

The amount of mechanical energy for vibration exposure depends on the amplitude of the vibrations, the frequency of the vibrations and the exposure time. According to the hypothesis P.A. Rebinder there is a limit value of the shear rate during vibration, causing the transition of concrete mixture from a state of elastic-plastic to a state of temporary fluidity. Upon reaching a certain impact energy, a transition to a state of temporary fluidity is observed. With the extension of the exposure time, the structural viscosity will decrease even more, but a further increase in shear will not cause a decrease in viscosity, but will lead to destruction of the structure of the cement mortar. Therefore, it is necessary to control the time and energy of the vibration effect on the cement mortar [11].

2.2.1. Laboratory tests. 1. Laboratory of SpetsUBR OJSC (Nizhnevartovsk, 1997). The spreadability index on the cone of the AzNII was: when using the VL-1 vibrator – more than 25 cm; when using a traditional mixer – 23 cm. The strength of cement stone increased by 8%. The experimental results depending on the processing time are presented in table 2. As a result of studies, it was determined that during vibration processing, the spreadability of the mortar increased, but during vibration processing for more than 10 minutes, the spreadability of the mortar decreased.

Table 2. Laboratory test results.

Cement mortar	Without vibration processing		Vibration processing time (min)		
	From cement aggregates	From averaging reservoir	10	20	30
Spreadability (cm)	18	23	25	21.5	19.5

2. Laboratory NIPImorneftegaz JV Vietsovpetro (Vung Tau, Socialist Republic of Vietnam, 1997). The cement mortar was prepared using a standard laboratory mixer and a laboratory vibrator VL-1 according to the formulation of the Scientific and Production Company Exbourg & Co. The spreadability of the solution was: when using a vibrator VL-1 – 25.5 cm; when using a standard mixer – 22 cm.

2.2.2. Field tests. NIPImorneftegaz JV Vietsovpetro (Vung Tau, Socialist Republic of Vietnam, 1997). Preparation of cement mortars in the process of completion of well No. 69 on the offshore stationary platform OSP-3, cement slurry LCH (lightweight cement for hot wells). The arrangement of mechanical mixers and VEMA is presented in Figure 4.

The density of the cement mortar is 1.52 g/cm^3. The spreadability of the mortar was: when using a VEMA vibrator – 27 cm; when using a standard mixer – 25 cm. It was established that due to the use of VEMA-0.3, stagnant zones disappeared in the averaging reservoir.

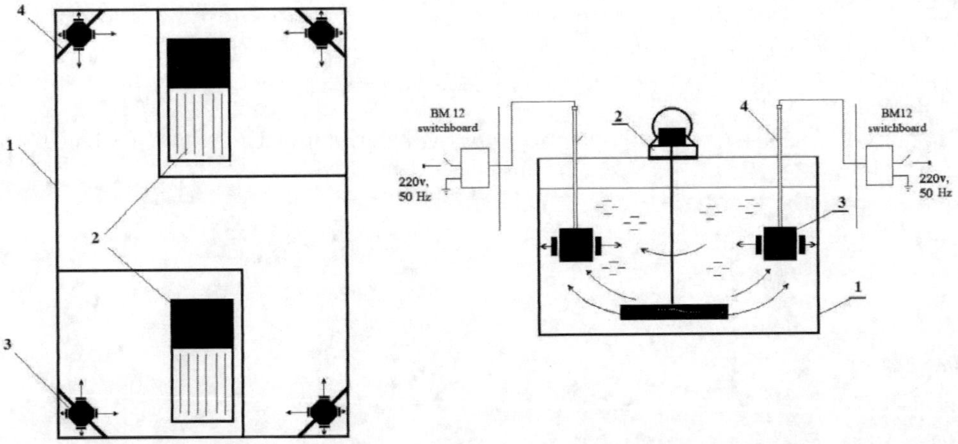

1 - averaging reservoir; 2 - mechanical mixers; 3 – VEMA; 4 - vibrator suspension fittings
Figure 4. Scheme of the experiment for the preparation of cement mortar at OSP-3 JV "Vietsovpetro".

3. Conclusion

1. The use of VEMA for the preparation of drilling mud significantly reduces the preparation time, reduces the energy intensity of the technological process and significantly reduces the ecotoxicity of flushing fluids.

2. The use of VJMA technology in the process of preparing cement mortars can increase the homogeneity and spreadability (decrease in viscosity) of the mortar while increasing the strength of the cement stone.

3. The time of vibration processing must be selected by the quality of the process fluid or by the amount of energy reported to the medium.

References

[1] Dmitrieva M, Leitsin V, Sharanova A and Tovpinets A 2018 *Matec Web of Conf.* **251** 01024

[2] Daneker V A and Rikkonen S V 2017 *Bulletin of the Tomsk Polytechnic University. Geo Assets Eng.* **328** 7 pp 86-92

[3] Daneker V A 2018 *Raschet i konstruirovaniye elektromagnitnykh preobrazovateley dlya aktivatsii zhidkikh sistem* (Tomsk: Tomsk Polytechnic University Press) p 102

[4] Azin A V, Bogdanov E P, Ponomarev S V and Rikkonen S V 2019 *Bulletin of the Tomsk Polytechnic University. Geo Assets Eng.* **330** 4 pp 41-48

[5] Tkachev O A and Tuguniv P I 1988 *Sokrashcheniye poter' nefti pri transporte i khranenii* (Moscow: Nedra Publishers) p 116

[6] Loskutova Yu V, Prozorova I V and Yudina N V 2011 *Chem. and Tech. of Fuels and Oils* **47** pp 358-361

[7] Volkova G I, Anufriev R V, Yudina N V and Tchaikovskaya O N 2016 *Russ. Phys. J.* **59** pp 1289-1294

[8] Volkova G I, Loskutova Yu V, Prozorova I V and Berezina E M 2015 *Podgotovka i transport problemnykh neftey: nauchno-prakticheskiye aspekty* (Tomsk: Tomsk Polytechnic University Press) p 134

[9] Chubik P S 1999 *Kvalimetriya burovykh promyvochnykh zhidkostey* (Tomsk: Scientific and technical literature publishing house) p 300

[10] JV Vietsovpetro 2001 *Development and coordination of the technical layout of the VSP-1.0 vibro-jet mixers with associated piping and their installation on the OSP for the preparation of cement and drilling muds. Report (final)* Contract No. LX 147 VSP GEOFIT 03.00. (Soviet Republic of Vietnam, Vung Tau)

[11] Rebinder P A 1979 *Izbrannyye trudy. Poverkhnostnyye yavleniya v dispersnykh sistemakh. Fiziko-khimicheskaya mekhanika* (Moscow: Nauka Publisher) p 388
[12] Logvinenko S V 1978 *Tekhnika i tekhnologiya tsementirovaniya skvazhin* (Moscow: Nedra Publishers) p 384
[13] Usmanov R A 2004 *Proc. of Young Scientists* (Ekaterinburg: Ural State Technical University) pp 33-36

IOP Conf. Series: Materials Science and Engineering **911** (2020) 012002 doi:10.1088/1757-899X/911/1/012002

Pore colmatation in case of liquid corrosion of concrete

S V Fedosov[1], V Eu Rumyantseva[2], V S Konovalova[2] and A S Evsyakov[2]
[1]Moscow State University of Civil Engineering (National Research University), department of technologies of cohesive materials and concretes, 129337, Yaroslavskoe av., 26, Moscow, Russian Federation
[2]Ivanovo State Polytecnic University, department of natural sciences and technosphere safety, Sheremetevsky av., 21, Ivanovo, 153000, Russian Federation

E-mail: fedosov-academic53@mail.ru, varrym@gmail.com, kotprotiv@yandex.ru, smart47@rambler.ru

Abstract. Pore colmatation plays a positive role, since a decrease in the permeability of cement concrete due to the deposition of insoluble corrosion products in the pores leads to a slowdown in corrosion processes. To test a mathematical model of pore colmatation of cement concretes under liquid corrosion, a study of mass transfer processes occurring when samples from Portland cement are exposed to a liquid aggressive medium was conducted. The article presents equations that allow us to determine the rate of pore colmatation and the thickness of the layer of colmatant formed in the pores during liquid corrosion of cement concretes. Based on the results of experimental studies of liquid corrosion of cement stone, the mass transfer parameters were determined in the investigated system «cement concrete – 2 % $MgCl_2$ solution». Based on the obtained data, the values of the colmatation rate and the thickness of the layer of colmatant of cement concrete were calculated during liquid corrosion in a 2 % $MgCl_2$ solution. It has been established that over time, the rate of mass transfer processes occurring in concrete decreases due to the difficulty of penetration of aggressive medium deep into concrete due to the deposition of corrosion products in the pores.

1. Introduction

Concrete is a capillary-porous material penetrated by a very fine network of pores and capillaries of various sizes [1]. When concrete is moistened, the smallest pores and capillaries are filled with an aggressive medium that reacts with calcium hydroxide to form insoluble reaction products that seem to clog these capillaries. Colmatation of pores and capillaries occurs, which leads to a decrease in the permeability of concrete [2, 3].

When cement stone is corroded, calcium hydroxide is dissolved in the outer layer of concrete [4, 5]. Calcium hydroxide is washed out of the concrete pores, while almost all the silica gel remains in the product pores, causing their partial clogging (colmatation) [5, 6]. Thus, the corrosion process becomes to some extent self-locking. The more silicic acid gel is formed during corrosion, the denser and less permeable it is for calcium ions, the more the process is inhibited over time [7].

2. Relevance

To assess the depth of corrosion damage, various equations are proposed for predicting the stability of concrete in an aggressive environment [8-12]. These equations take into account the rate of concrete corrosion in the initial period, the rate of heterogeneous reactions and the nature of control (kinetic or diffusion), the kinetics of the internal diffusion process with a constant diffusion coefficient over time, the porosity of the cement stone, the dependence of mass transfer coefficients on the structure and composition of concrete, as well as the composition and concentration of aggressive media, relative humidity (saturation level) and the concentration of calcium ions in the pore solution, the formation of calcite and the dissolution of hydration products, simultaneous impact of the force load and negative impact of aggressive media on concrete and reinforced concrete structures.

Based on the theory of mass transfer processes, equations are proposed for the mathematical description of the kinetic dependences of the corrosion process, which are most often encountered in practice. These equations describe extensive and intensive processes of concrete corrosion inhibition [13-16]. However, the developed models do not fully take into account the process of colmatation of the pores and capillaries of concrete, which remains completely unexplored.

3. Problem statement

Since pore colmatation influences the development of mass transfer processes and the rate of concrete corrosion, it is necessary to conduct research on the factors affecting the rate of colmatant deposition in the pores and the spread of the colmatation front. Establishing the regularities of mass transfer processes in the case of liquid corrosion of concrete, taking into account the pore colmatation, will allow using natural and artificial colmatation to prevent the development of corrosion damage.

4. Results of experimental studies

For the exponential dependence of the change in the mass conductivity coefficient over time, the expressions for determining the colmatation rate $\frac{dL(Fo_m)}{dFo_m}$ and the thickness of the colmatant layer $L(Fo_m)$ have the form [17]:

$$\frac{dL(Fo_m)}{dFo_m} = A_1 \cdot exp(-\hat{B}_1 Fo_m) \cdot \frac{\partial\theta_B(\bar{R}_k, Fo_m)}{\partial\bar{x}} \cdot \frac{\nu_{AD} \cdot K_{\rho c}}{(1-\varepsilon_{sed})}, \quad (1)$$

$$L(Fo_m) = \frac{\partial\theta_B(\bar{R}_k, Fo_m)}{\partial\bar{x}} \cdot \frac{A_1 \cdot K_{\rho c} \cdot \nu_{AD}}{\hat{B}_1 \cdot (1-\varepsilon_{sed})} \cdot [1 - exp(-\hat{B}_1 Fo_m)]. \quad (2)$$

Здесь: $\hat{B}_1 = \frac{B_1 \cdot \delta_{con}^2}{k_{B0}}$; A_1, B_1 are experimentally determined exponent coefficients; δ_{con} is the thickness of the concrete layer, m; k_B is coefficient of mass conductivity of the transferred component of calcium hydroxide in concrete, m^2/s; $Fo_m = \frac{k_B \cdot \tau}{\delta_{con}^2}$ is Fourier mass transfer criterion; $\theta_B = \frac{C_{B.o.} - C_B(x,\tau)}{C_{B.o.}}$ is dimensionless concentration; $C(x,\tau)$ is the concentration of «free $Ca(OH)_2$» in concrete at the time τ at an arbitrary point with the x coordinate, in terms of CaO (kg CaO/kg of concrete); C_0 is concentration of «free $Ca(OH)_2$» in concrete at the initial moment of time at an arbitrary point with the x coordinate, in terms of CaO (kg CaO/kg of concrete); $\frac{\partial\theta(\bar{R}_K, Fo_m)}{\partial\bar{x}}$ is mass flow of the transferred component; ν_{AD} is stoichiometric coefficient depending on the reaction; ε_{sed} is porosity of the sediment layer, which characterizes the volume of voids in the sediment layer; $K_{\rho c} = K_\rho \cdot C_{B.o.}$ is density coefficient.

For Figure 1 the profiles of the dependence of the colmatation rate, $\frac{dL(Fo_m)}{dFo_m}$, and the thickness of the colmatant layer, $L(Fo_m)$, on the Fourier mass transfer criterion, Fo_m, for the porosity of the sediment layer $\varepsilon_{sed} = 0{,}5$, for the changing value of the mass flow, $\frac{\partial\theta(\bar{R}_K, Fo_m)}{\partial\bar{x}}$, are presented.

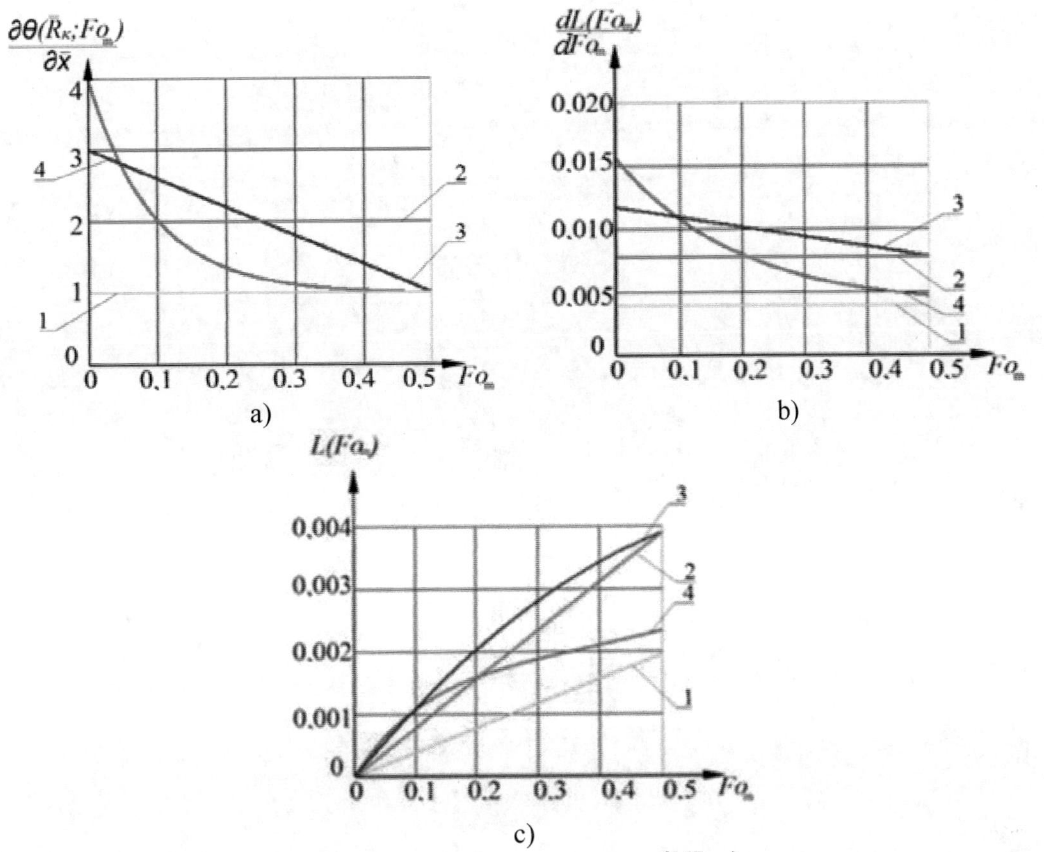

Figure 1. The dependence of the colmatation rate (b), $\frac{dL(Fo_m)}{dFo_m}$, and the thickness of the colmatant layer (c), $L(Fo_m)$, on the Fourier mass transfer criterion, Fo_m, for the porosity of the sediment layer ε_{sed} = 0,5, for the value of the mass flow (a), $\frac{\partial\theta(\bar{R}_K,Fo_m)}{\partial\bar{x}}$: 1,2 – constant; 3 – changing along an inclined linear relationship; 4 – changing exponentially.

The results of the research presented in Figure 2 show that after 70 days of testing, the concentration of Ca^{2+} cations in the liquid phase reaches constant values, and after 150 days, a state close to equilibrium occurs in the system under study [18, 19]. The values of mass transfer characteristics were calculated using a mathematical model of liquid corrosion of the second type of cement concretes [19, 20].

Figure 3 shows that over time, the coefficient of mass conductivity decreases according to an exponential law, which corresponds to a deceleration of the mass transfer processes occurring in concrete during liquid corrosion due to the formation of corrosion products in the pores.

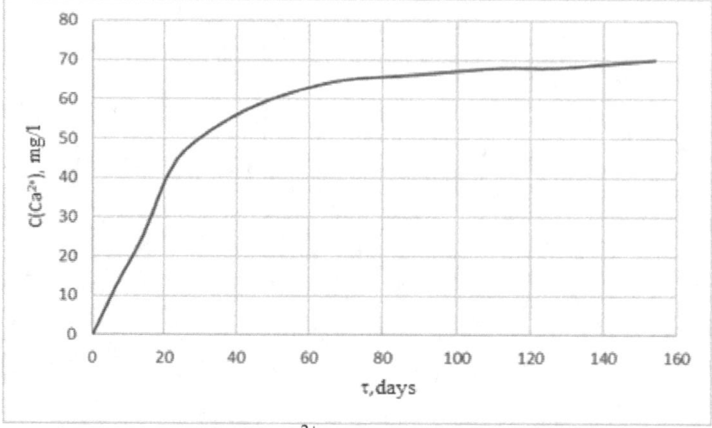

Figure 2. Kinetic curve of Ca^{2+} cations in a 2 % $MgCl_2$ solution for a cement concrete sample.

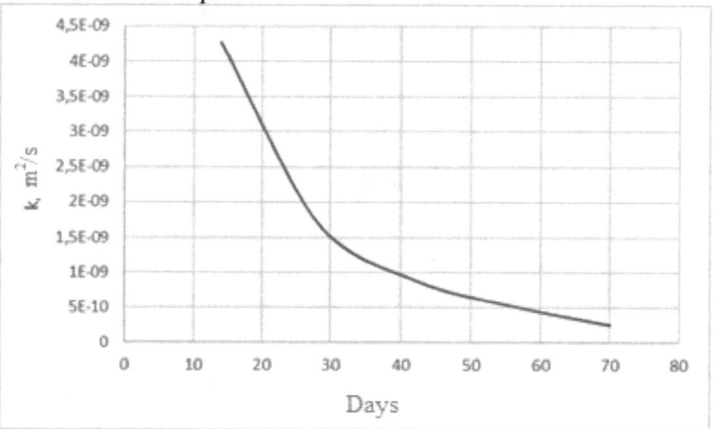

Figure 3. Change in the value of the coefficient of mass conductivity in the system «cement concrete – 2 % solution of $MgCl_2$» over time.

Using equations (1) and (2), the calculation of the rate of colmatation of cement concrete without additives and the thickness of the colmatant layer under liquid corrosion in 2 % $MgCl_2$ solution was performed (Figures 4 and 5).

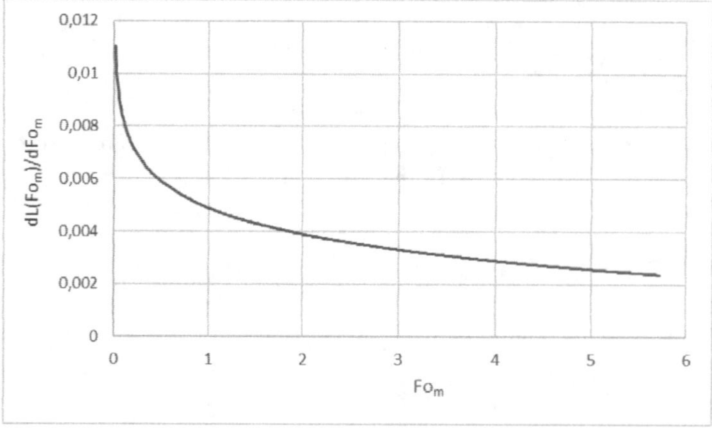

IOP Conf. Series: Materials Science and Engineering **911** (2020) 012002 doi:10.1088/1757-899X/911/1/012002

Figure 4. Change in time of colmatation rate $\frac{dL(Fo_m)}{dFo_m}$ for liquid corrosion in 2 % $MgCl_2$ cement concrete solution.

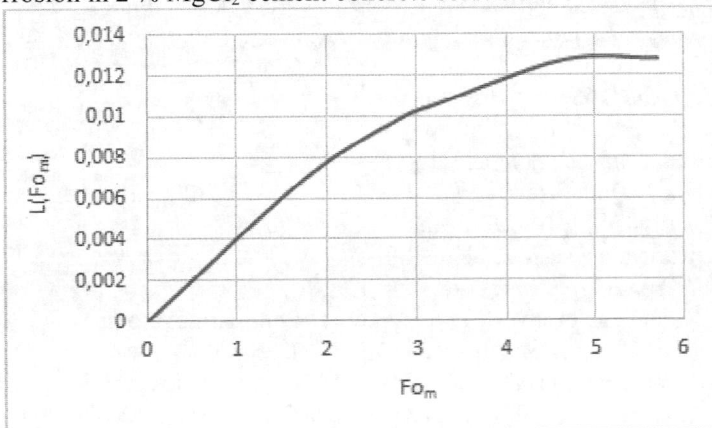

Figure 5. Change in time of the thickness of the layer of colmatant $L(Fo_m)$ during liquid corrosion in a 2 % solution of $MgCl_2$ cement concrete.

The form of the obtained curves corresponds to the dependencies calculated using the mathematical model of colmatation, shown in Figure 1 (curve 4). The established values of the rate of colmatation of cement concrete under liquid corrosion in 2 % $MgCl_2$ solution and the thickness of the colmatant layer correspond to the values predicted by mathematical modeling, which allows us to judge the adequacy of the developed model of pore colmatation of cement concretes.

5. Conclusion

Having an understanding of the kinetics of mass transfer processes and the mechanism of natural colmatation during liquid corrosion of cement concretes, it is possible not only to predict the durability of concrete, but also to control the processes of corrosion destruction, affecting the rate of corrosion by artificial colmatation of the pores and capillaries of concrete.

References

[1] Mehta P K and Monteiro P J M 2006 *Concrete: Microstructure, Properties, and Materials* (New York: McGraw-Hill) p 683

[2] Yanakhmetov M R, Chuykin A E and Massalimov I A 2015 *Nanotechnologies in construction: a scientific internet-journal* **7 1** pp 63-72

[3] Mukhametshin V V and Kadyrov R R 2017 *Nanotechnologies in construction: a scientific internet-journal* **9 6** pp 18-36

[4] Stepanova V F, Rozental N K, Chekhniy G V and Baev S M 2018 *Construction materials* **8** pp 69-73

[5] Nikolopoulos C V 2015 *Journal of engineering mathematics* **91 1** pp 143-163

[6] Bazhenov Y M, Erofeev V T, Rimshin V I, Markov S V and Kurbatov V I 2016 *Engineering Solid Mechanics* **4 4** pp 219-225.

[7] Qiao H, Shi Y, Chen D, Cao H and Gao S 2015 *Chongqing Daxue Xuebao* **38 6** pp 129-137

[8] Levandovskiy A N, Melnikov B E and Shamkin A A 2017 *Magazine of Civil Engineering* **1 (69)** pp 3-22

[9] Sugiyama T, Ritthichauy W and Tsuji Y 2008 *Cement and Concrete Research* **38** pp 49-67

[10] Ovchinnikov I I, Snezhkina O V and Ovchinnikov I G 2017 *IOP Conference series: Materials and engineering* **262** 012041
[11] Leonovich S N, Shalyi E E and Kim L V 2019 *Science and Technique* **18 4** pp 284-291
[12] Gusev B V and Fayvusovich A S 2019 *Industrial and civil engineering* **7** pp 58-63
[13] Patel R A, Perko J, Jacques D, Schutter G D, Ye G and van Breugel K 2018 *Construction and Building Materials* **165** pp 833-845
[14] Gilmutdinov T Z, Fedorov P A, Latypov V M, Lutsyk E V and Latypova T V 2017 *Journal of Engineering and Applied Sciences* **12 15** pp 4406-4413
[15] Hoang N D, Chen C T and Liao K W 2017 *Measurement* **112** pp 141-149
[16] Isgor O B and Weiss W J 2019 *Materials and structures* **52** 3
[17] Fedosov S V, Rumyantseva V E, Krasilnikov I V, Konovalova V S and Evsyakov A S 2018 *Magazine of Civil Engineering* **7 (83)** pp 198-207
[18] Fedosov S V, Roumyantseva V E, Krasilnikov I V and Konovalova V S 2018 *IOP Conference Series: Materials Science and Engineering* **456** 012039
[19] Fedosov S V, Roumyantseva V E, Krasilnikov I V, Konovalova V S and Evsyakov A S 2018 *IOP Conference Series: Materials Science and Engineering* **463** 4 042048
[20] Fedosov S V, Roumyantseva V E, Konovalova V S and Loginova S A 2018 *International Journal for Computational Civil and Structural Engineering* **14 3** pp 161-168

IOP Conf. Series: Materials Science and Engineering **911** (2020) 012003 doi:10.1088/1757-899X/911/1/012003

Cement based foam concrete with hardening accelerators

A B Steshenko[1], A I Kudyakov[1] and N E Ryabtseva[1]
[1]Tomsk State University of Architecture and Building, Department of Building materials, 2 Solyanaya Sq., 634003, Tomsk, Russian Federation

E-mail: steshenko.alexey@gmail.com

Abstract. The results of studies of the effect of hardening accelerators on the properties of cement based foam concrete mix and foam concrete of natural hardening are given in this article. The study was conducted in the Laboratory of Tomsk State University of Architecture and Building which is accredited in accordance with the national standards requirements. It has been proposed to decrease the shrinkage deformations of the concrete mix and increase compressive strength of hardened foam concrete by affecting the cement matrix with the calcium chloride and calcium oxalate in the amount of 2 and 0.5% of the cement mass, respectively. This increases the aggregative stability and reduces the plastic shrinkage of the foam concrete mix by 47-65%. In foam concrete with the hardening accelerators the compressive strength rises from 0.31 to 0.47 MPa and from 0.96 to 1.44 MPa at the age of 28 days while maintaining the average density of D400 and D600 respectively. The developed composition of cement based foam concrete is recommended for constructing monolithic walls at a construction site or wall blocks in factory conditions for individual housing construction.

1. Introduction

Every year the building materials market expands and expands its range with new types. This contributes to the continuous development of production technologies. This factor has a positive effect on the quality of the material produced. In the conditions of the economic crisis, consumers are guided by material that allows them to save resources, both at the stage of construction of individual housing, and during its operation. Therefore, the construction of foam concrete blocks of natural hardening has gained particular relevance [1, 2]. The low thermal conductivity and energy intensity of the foam concrete wall blocks provide energy savings during the operation and construction of buildings and structures [1-5]. Shrinkage deformations, which adversely affect the quality of foam concrete, can be attributed to negative qualities. Shrinkage deformations lead to the formation of cracks in the products, which leads to a decrease in strength and an increase in the thermal conductivity of foam concrete. This fact greatly reduces the effectiveness of its use in wall structures of buildings [4-7].

Factors affecting the shrinkage deformation of foam concrete are as follows: hardening conditions (normal conditions, dry hardening, steam curing), foaming agent properties (foam stability, foam expansion), cement properties (w/c, type of cement), sand properties (content of harmful impurities, particle-size distribution), manufacturing process and equipment (time of aeration and mixing intensity of the mixture, sequence of loading components). Also one of the main factor affecting the quality of foam concrete wall blocks is the correctly selected technology for their manufacture. Today, there are many ways of preparing foam concrete, which have their drawbacks and advantages, which are difficult to accurately determine due to the wide variety of foam concrete mixtures production and the small amount of statistical data on the properties of the resulting foam concrete. Based on previous studies and a literature review, the authors recommend the use of a one-stage technology for the preparation of foam concrete [5, 7].

The following shrinkage deformations are distinguished depending on the time of structure formation of foam concrete:

— contraction – is the result of the chemical interaction of concrete components with water and does not only change the external dimensions of the sample, but rather contributes to changes in the pore structure of the material: the volume of pores occupied by water decreases, air pores arise. Usually, this shrinkage develops during the hardening of concrete, when it is still sufficiently plastic, and therefore not accompanied by noticeable cracking of the material [8];

— drying shrinkage – is associated with a decrease in the moisture content of concrete, that is, with the evaporation of free water in a hardened cement paste and is caused by capillary phenomena. Drying shrinkage can lead to cracking in conditions of elevated temperature, as well as in hot climates, when concrete does not receive the necessary moisture care [3];

— plastic shrinkage – shrinkage of a freshly laid compacted concrete mixture. It develops within the first 3 hours from the moment of laying the foam concrete mixture and depends on the aggregate stability of the foam concrete mixture [9];

— carbonization shrinkage – occurs as a result of the reaction of portlandite contained in a cement stone with carbon dioxide to form calcium carbonate $CaCO_3$. The total volume of foam concrete is reduced, the structure is destroyed, and the strength is reduced. The carbonization of foam concrete is influenced by the concentration of CO_2 in the air, humidity and air temperature, type of cement, water-cement ratio, hardening conditions and maintenance.

The problem of controlling the processes of foam concretes production with low shrinkage is very relevant and is still not fully resolved [4-7, 10, 11].

The following technological methods have found their application to regulate the shrinkage of foam concrete, the setting time and the speed of hardening of binders.

— use of reinforcing fibers. Most researchers note [1, 4, 7, 12] that an effective solution to the problem is the use of various types of reinforcing fibers that can absorb tensile stresses throughout the entire volume of the product. Shrinkage reduction is achieved by introducing fiber into the composition of cellular concrete. An effective technique is polydisperse reinforcement, which involves the use of a combination of fibers with different characteristics. Shrinkage deformations of fiber-reinforced concrete depend on the elastic modulus of cellular concrete and fibers, as well as on the fraction of volume reinforcement, coefficient of volume reinforcement;

— the use of active mineral admixture. Ground quartz sand or other types of siliceous materials are used for autoclaved foam concrete: marshalite, tripoli, diatomite, flask, loess and others, as well as metallurgical granulated slags, fly ash of a power plant, including the use of aluminosilicate microspheres [13, 14, 15];

— the use of modifying additives. The most effective technological technique for regulating the shrinkage of foam concrete is the use of modifying additives, in particular hardening accelerators [4, 6, 16]. The introduction of hardening accelerator additives in the foam concrete mixture, even in small amounts, provides accelerated initial structure formation of the foam concrete mixture, thereby significantly reducing the shrinkage deformation of foam concrete. The most common and most effective hardening accelerators are chlorides and mixtures based on them.

The aim of this study is to establish patterns of influence of hardening accelerators on the properties of foam concrete mix and foam concrete.

2. Materials and methods

In carrying out experimental research were used raw materials that meet the requirements of national standards of Russia. As binders Portland cement of Topki Cement Factory (Kemerovo Region, Russia) CEM I 42.5H (Russian State Standard GOST 30515-2013) was used, sand of Kudrovskoe deposit of Tomsk Region (Russia) with fineness modulus 1.7 (Russian State Standards GOST 8736-2014 and GOST 26633-2012) and water (Russian State Standard GOST 23732-2011).

IOP Conf. Series: Materials Science and Engineering **911** (2020) 012003 doi:10.1088/1757-899X/911/1/012003

PB-Lux (Specifications 2481-004-59586231-2005) was chosen as foam agent based on the results of the evaluation of the quality of the foam agents [4].

The calcium chloride (Russian State Standard GOST 450–77 "Technical calcium chloride. Technical conditions") and calcium oxalate were used as a hardening accelerators. Calcium oxalate is a salt of an alkaline-earth metal of calcium and organic dibasic oxalic acid. Hardening accelerators were introduced in the following dosage: calcium chloride 2%, calcium oxalate 0.5% by weight of cement.

Physical and mechanical properties of foam concrete were defined in accordance with Russian State Standards requirements: compressive strength (GOST 10180-2012); average density (GOST 12730.1-78). Plastic shrinkage and plastic strength were determined within the first three hours from the moment of foam concrete mixture placement into the mould according to methodology described in [4].

Compositions of foam concrete were selected in accordance with Russian Standard SN 277-80 "Instructions for the production of cellular concrete products" and given taking into account the actual average density of the concrete mix (Table 1) [4-6].

Table 1. Compositions of foam concrete per 1 m^3

Type of foam concrete	Components consumption					
	cement, kg	sand, kg	water, l	foam agent, l	hardening accelerator, kg	Average density grade
Reference No.1	210	105	157.5	1.7	-	D400
With calcium oxalate 0.5%	210	105	157.5	1.7	1.05	D400
Reference No.2	280	140	210	1.5	-	D600
With calcium chloride 2%	280	140	210	1.5	5.6	D600

Foam concrete mixture formation was performed in one stage using laboratory foam concrete mixer [4]. The molded foam concrete samples are marked and maintained under normal conditions at a temperature of 20±2°C for a day, after which they are released and placed in a normal hardening chamber, where they are stored for 7 and 28 days until testing at a temperature of 20±2°C and relative humidity of not less than 90%. Testing and evaluation of the quality of foam concrete was carried out according to Russian State Standard GOST 25485–89.

3. Results and discussions

Plastic strength is an important parameter of the rheological characteristics of the foam concrete mix. Inter-pore partitions should provide preserving the cellular structure when the elasticity of the surfactant films decreases so that it is insufficient to hold the gas phase. It is known from practice that the inter-pore partitions of foam mixtures acquire the ability to fix the geometry of the gas phase only after a sufficient number of viscous contacts between the particles of the solid phase are replaced by elastic ones. The time required for the formation of such contacts in cement-containing disperse systems is from 150 minutes or more [4]. Based on the foregoing, the test of plastic strength lasted 3 hours. The results of the effect of hardening accelerators on the plastic strength of the foam concrete mixture are presented in Figure1.

As the results shown in Figure 1, foam concrete mixtures with additives in the initial stages of hardening have higher plastic strength values in comparison with foam concrete mixture without additives.

The plastic strength of foam concrete mixtures with additives is 2-3 times greater in comparison with the control composition after 3 hours of hardening. With the introduction of calcium oxalate in an amount of 0.5% by weight of cement leads to insreas R_{pl} from 128.2 to 418.7 Pa, and using calcium chloride in an amount of 2% by weight of cement leads to increase R_{pl} from 256.4 to 535.1 Pa.

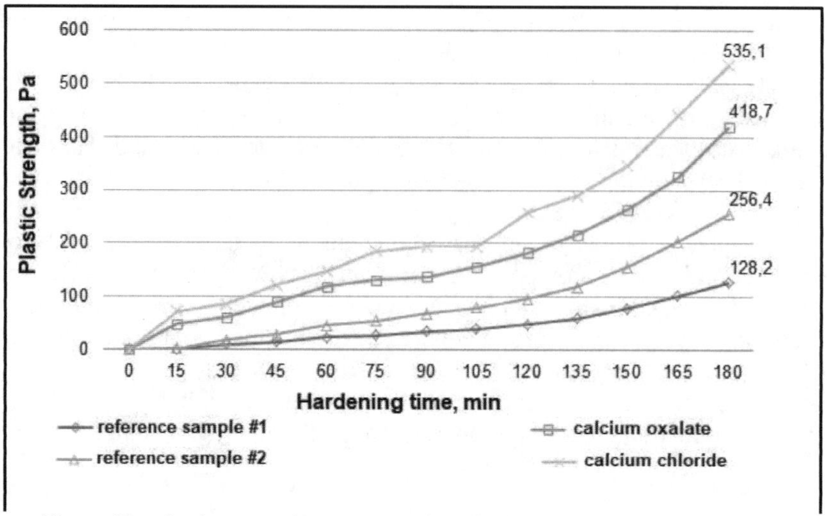

Figure 1. The effect of hardening accelerators on the plastic strength of the foam concrete mixture

Based on the results presented in fig. 1 it is shown that the introduction of hardening accelerators into the foam concrete mixture allows it possible to accelerate the initial structure formation of foam concrete and increase the aggregative stability of foam concrete mixtures [4, 6, 8]. It can be concluded that a consequence of an increase in the aggregate stability of foam concrete structures from the beginning to the end of cement setting will be a decrease in the shrinkage deformations of the foam concrete mixture and foam concrete. In this regard, studies have been conducted on the effect of hardening accelerators on the shrinkage deformation of the foam concrete mixture.

Figure 2 shows the results of a study of the effect of additives on the plastic shrinkage (ε_{pl}) of foam concrete mixtures. It was found that the addition of hardening accelerators to the foam concrete mixture leads to decrease the plastic shrinkage by 47–65%. Using calcium chloride allows to decrease plastic shrinkage by 65% compare with the results of testing the foam concrete mixture without additives (reference No.2), using calcium oxalate - by 47%, compared with the results of testing the foam concrete without additives (reference No.1). The obtained data confirms the results of a study on the effect of hardening accelerators on the process of early structure formation of foam concrete. Accelerating this process, hardening accelerators stabilize the porous structure of the foam concrete mixture and thereby reduce the shrinkage deformation of the foam concrete mixture.

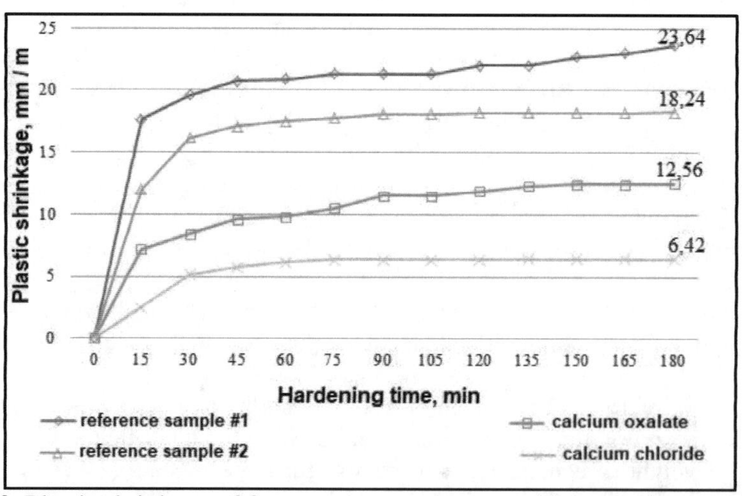

Figure 2. Plastic shrinkage of foam concrete mixtures with hardening accelerators

Since the properties of the foam concrete mixture determine the quality parameters of foam concrete, the following studies have been conducted on the effect of hardening accelerators on the properties of hardened foam concrete.

The main physical and mechanical properties of foam concrete are strength characteristics and density. The density and strength of foam concrete are closely related, the higher the density, the higher the strength. In Figure 3 presents the results of studies to determine the values of compressive strength and average density of foam concrete samples with and without hardening accelerators.

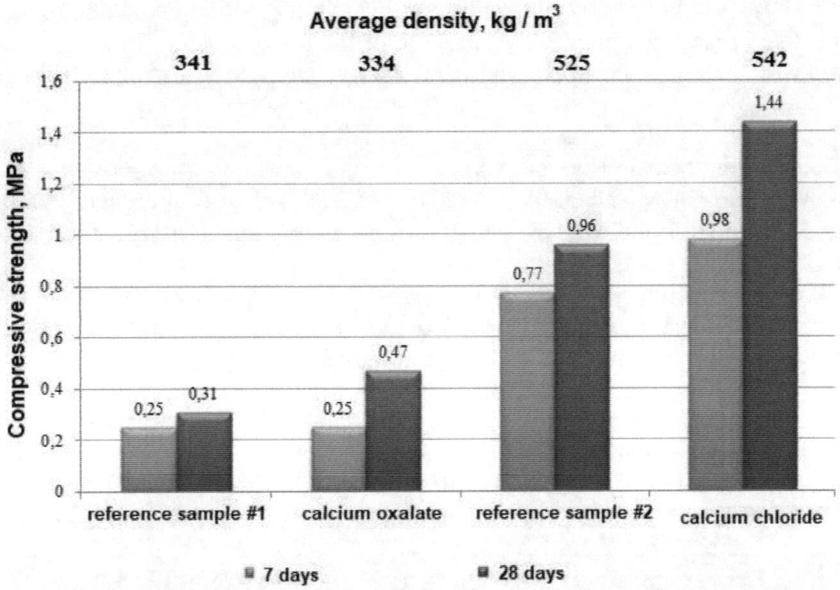

Figure 3. The effect of hardening accelerators on the properties of foam concrete

It is shown from Figure 3 that using calcium chloride in an amount of 2% by weight of cement allows to increase the compressive strength of foam concrete at 7 days age from 0.77 to 0.98 MPa. The introduction of calcium oxalate in an amount of 0.5% by weight of cement leads to increase the compressive strength of foam concrete by 51.6% in comparison with the reference No.1 samples at 28 days age, the density varies slightly. In the same terms, with the introduction of calcium chloride in an amount of 2% by weight of cement, the compressive strength of foam concrete increases by 50% compared with the the reference No.2 samples, while the density also changes slightly.

Calcium chloride increases the solubility of free lime released by cement, and accelerates the hydration of clinker minerals. As a result, the hydration of cement increases in the initial period, due to that the formation of gel increases, and the early strength of foam concrete increases.

Calcium oxalate increases the strength of concrete by accelerating the hydration of the main strength carrier - tricalcium silicate.

4. Conclusion

An effective heat-insulating and structural-heat-insulating material is foam concrete. The mechanism for reducing shrinkage deformations of foam concrete and foam concrete mixture is based on the addition of hardening accelerators, which reduce shrinkage and cracking due to the accelerated process of initial structure formation of natural hardening foam concrete.

According to the results of the study, it was found that:

1. The introduction of calcium chloride in an amount of 2% by weight of cement into the foam concrete mixture leads to increase the plastic strength of the foam concrete mixture after three hours of hardening from 256.4 to 535.1 Pa, while the plastic shrinkage decreases by 64.8%, compared to test

results of foam concrete mixture without additives. When calcium oxalate is introduced into the foam concrete mixture in an amount of 0.5% by weight of cement, the plastic strength of the foam concrete mixture after three hours of hardening increases from 128.2 to 418.7 Pa, while the plastic shrinkage decreases by 47.7%, compared to reference No.2.

2. The introduction of hardening accelerators leads to an increase in the compressive strength of foam concrete in the early stages of hardening. The introduction of calcium oxalate in an amount of 0.5% by weight of cement allows to increase the compressive strength at 28 days age from 0,31 to 0,47 MPa (by 51.6%), compared with the reference No.1, while the average the density slightly decreases from 341 to 334 kg/m^3. The introduction of calcium chloride in an amount of 2% by weight of cement leads to increase the compressive strength of the foam concrete from 0.96 to 1.44 MPa (by 50%) compared with the reference No.2, while the density increased from 525 to 542 kg/m^3.

References

[1] Dunn T van Zijl and G van Rooyen A S 2018 *Journal of Building Engineering* **20** pp 663–670
[2] Vytchikov Y, Saparev M and Golikov V 2018 *MATEC Web of Conferences* **196** 04015
[3] Amritha R, Dhanya S and Mini K M 2019 *Construction and Building Materials* **221** pp 787–799
[4] Steshenko A B and Kudyakov A I 2015 *Magazine of Civil Engineering* **2** pp 56–62
[5] Steshenko A B, Kudyakov A I 2018 *Magazine of Civil Engineering* **84** (8) pp 86–96
[6] Kudyakov A I and Steshenko A B 2015 *Lett. Mater* **5** (1) pp 3-6
[7] Kudyakov A I and Steshenko A B 2015 *IOP Conf. Ser.: Mater. Sci. Eng.* **71** 012019
[8] Ghourchian S, Wyrzykowski and Markus G 2011 Measuring the early shrinkage of mortars drymix mortar *Yearbook ed Ferdinand Leopolder* **11-16**
[9] Saliba, J, Rozière E, Grondin F and Loukili A 2011 *Cem. Concr. Res.* **33 (2)** pp 209–217
[10] Alireza Bagheri S and Samea A *Journal of Materials in Civil Engineering* **30(6)** 04018091
[11] Namsonea E, Šahmenkoa G and Korjakin A 2017 *Procedia Engineering* **172** pp 760–767
[12] Yakovlev G, Keriene J, Gailius A and Girniene I 2006 *Mater. Sci.* **12 (02)** pp 147–151
[13] Kudyakov A, Prishepa I and Tolchennikov M 2015 *IOP Conf. Ser.: Mater. Sci. Eng.* **71** 012016
[14] Gökçe H S, Hatungimana D and Ramyar K. 2019 *Constr. Build. Mater.* **194** pp 1–11
[15] Abdulrahman Hilal, Thom A and Dawson N 2015 *International Journal of Engineering and Technology* **7 (4)** pp 286–293
[16] Kudyakov A I, Steshenko A B, Simakova A S and Latypov A D 2019 *IOP Conf. Series: Materials Science and Engineering* **597** 012037

IOP Conf. Series: Materials Science and Engineering **911** (2020) 012004 doi:10.1088/1757-899X/911/1/012004

Approach to evaluating the change of properties of the geosynthetic material used to stabilize the marine landscape slopes

V N Leitsin[1], A O Tovpinets[1], B V Chubarenko[2], D A Domnin[2], E E Esiukova[2] and E M Burnashov[3]

[1]Immanuel Kant Baltic Federal University, 14 Nevskiy St, Kaliningrad, 236016, Russian Federation
[2]Shirshov Institute of Oceanology of Russian Academy of Sciences, 36 Nakhimovsky Prospekt, Moscow, 117997, Russian Federation
[3]State budget institution of the Kaliningrad region "Baltberegozashchita", 1 Khutorskaya St., Svetlogorsk, 238560, Kaliningrad region, Russian Federation

E-mail: leitsin@mail.ru

Abstract. Changes in the properties of geosynthetic materials (geomats) used in coastal protection structures that protect the high coast from erosion, landslide and talus processes are analyzed. The proposed approach includes the selection of a representative set of sampling points for geomats, which takes into account various exposure conditions, conditions for the extraction of groundwater or in dry places, under the direct influence of the wind or in continuous grass cover, and other local operating features. In the process of cameral work with samples of the geomat, taken at the sampling points, it was proposed to assess the degree of degradation of the geomat, used to keep the slope from breaking, by the value of the breaking load of individual polymer filaments, while assessing the degree of microdamage of the material under study. Objective monitoring of microdamage of filaments is carried out by computed X-ray tomography. Using two geomat sampling as an example, the effectiveness of the approach is shown. Regular monitoring of the properties of geosynthetic material will provide an estimate of the residual life of the shore protection structure, a selection of recommendations for its repair and optimization.

1. Introduction

The beaches of the Kaliningrad region (Russia) are subject to pollution by geosynthetic materials used in coastal protection structures [1, 2]. The complex of engineering measures for strengthening the slopes (CESS) in Svetlogorsk is the largest shore protection structure on the coast of the Kaliningrad Region, the construction of which used geosynthetic material. It was built in 2015-2017 and is currently in operation (Figure 1). The area of the fortified surface is 90 thousand m^2 [3]. CESS performs the function of protecting the promenade from erosion, landslide and talus processes.

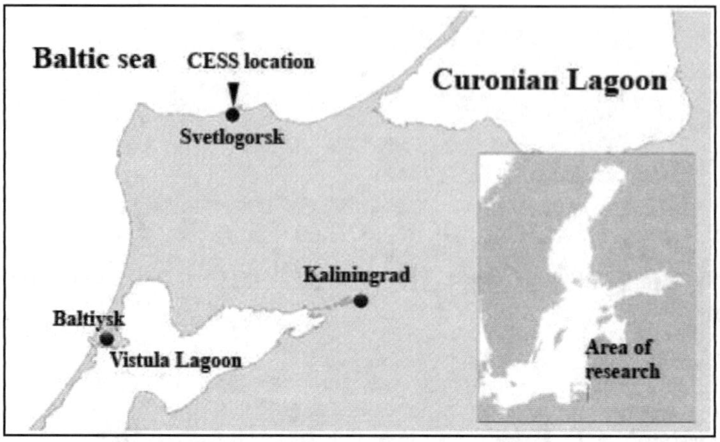

Figure 1. Location of the complex of engineering measures for strengthening the slopes (CESS) in Svetlogorsk

Geomat – a geosynthetic material in contact with air – is the main anti-erosion protection on landslide slopes, where stable slopes have a steepness of 1:1.5-1:2. This material is held by a metal mesh and covers the entire surface of the slope. Slopes with a steepness of more than 1: 1.5 are secured by arranging a reinforced embankment. To protect against erosion in the middle and upper parts of the slope, where the steepness reaches 50°, erosion protection was performed with sowing of perennial grasses, and the slope was retained using self-drilling anchors and a metal mesh.

In general, the slope has a northwestern exposure. For this shore protection structure, various types of geosynthetic materials were used. This is, in fact, the first experiment for the Kaliningrad region, where all artificial (geosynthetic) materials are stacked in such a way that they do not damage, but bypass existing vegetation. Thus, in the process of exploitation and natural environmental conditions, geosynthetic materials are gradually hidden by natural vegetation.

In reality, the geosynthetic material used is in different conditions: under the influence of direct solar radiation; in the shadow; in conditions of draining groundwater or in dry places where even rain moisture is not retained; under the direct influence of the wind or in continuous grass cover.

The aim of the work is to develop an approach to assessing the state of geosynthetic material used in coastal protection structures that protect landslide slopes from erosion, landslide and talus processes.

2. Materials and methods

To determine the degree of degradation of the main material used to hold the landslide slope from destruction – a plastic geomat – in the period 2018-2019, samples were taken of individual parts of the geomat, located in different exposure conditions. Several separate sections of the complex of engineering measures for strengthening the slopes were selected as test sites located east of the main descent to the sea in Svetlogorsk (near the Estrada Theater).

In total, 9 sampling points were determined at which geomaterial samples were taken at different time periods. Five points are located in the upper part of the slope and four points at its foot at heights of approximately 25 and 5 m, respectively (Figure 2).

IOP Conf. Series: Materials Science and Engineering **911** (2020) 012004 doi:10.1088/1757-899X/911/1/012004

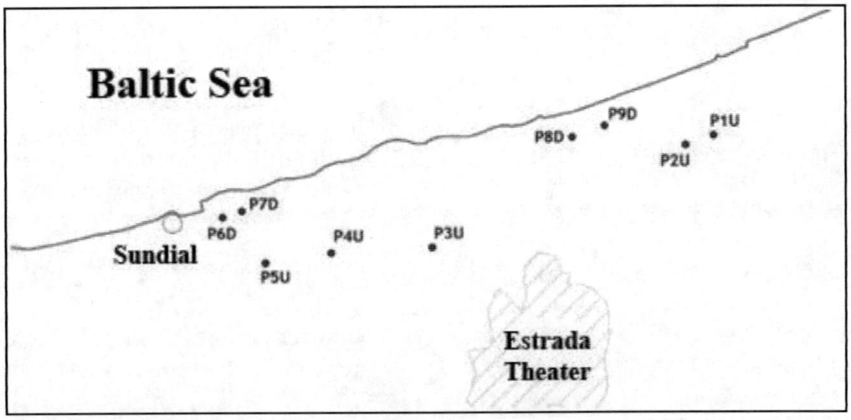

Figure 2. Sampling points for geomats at the complex of engineering measures for strengthening the slopes in Svetlogorsk. For the border of the depicted coastline adopted the outer border of the promenade.

Table 1 provides information on sampling points and operating conditions, including ultraviolet radiation. The point identifier contains a serial number (from 1 to 9), the letter index corresponds to the position of the point (U is the top of the slope, D is the foot of the slope, and the suffix summarizes the operating conditions (Sun - the sample was in an open sunny area, Shd - the sample was in the shady area, Grass - the sample was covered with grass.) Over the entire period of time, 34 samples were taken at all points.

3. Research and experiment

Instron ElectroPuls E1000 is used to conduct studies of the extreme mechanical characteristics of the geomat. A feature of this test machine is the ability to set loading modes in efforts (the value of the load determined by the sensor can control the movements of the actuator so that the specified mode of forces is realized).

Table 1. Sampling points for the geomat used for the complex of engineering measures for strengthening the slopes in Svetlogorsk

Point number	Latitude (degree, N)	Longitude (degree, E)	Description of conditions
P1U_Sun	54.947211	20.161953	The upper part of the slope in the illuminated area, the fragment is not covered with a steel mesh, the sample was taken from the edge, in the sun, not far from the tree, everything is covered with rot
P1U_Sun-Grass	54.947211	20.161953	The upper part of the slope, the fragment is not covered with a steel mesh, the sample was taken from the edge, in the grass, partial shade from the tree
P2U_Sun-Grass	54.947131	20.161564	The upper part of the slope, the fragment is not covered with a steel mesh, the sample was taken at the edge of the slope in the sun
P3U_Shd	54.9463	20.158081	The upper part of the slope, covered with steel mesh, in the shade
P4U_Sun	54.946244	20.156722	The upper part of the slope, the fragment is

			not covered with a steel mesh, the sample was taken from the edge of the slope in the sun
P5U_Shd	54.946162	20.155824	The upper part of the slope, covered with steel mesh, the sample was taken in the shade, with clay and rot.
P6D_Shd	54.946516	20.155227	The lower part of the slope, covered with a steel mesh, the sample was taken in the shade
P7D_Shd	54.946571	20.155495	The lower part of the slope, covered with a steel mesh, the sample was taken in the shade
P8D_Sun	54.947183	20.159987	The lower part of the slope, not covered with a steel mesh, the sample was taken in the sun
P9D_Sun	54.947278	20.160439	The lower part of the slope, covered with a steel mesh, the sample was taken in the sun

Samples were selected and prepared in random order from a geomat taken at two different points of the investigated CESS, characterized by significantly different operating conditions.

Figure 3 shows an example of a geomat sample used before and after the test.

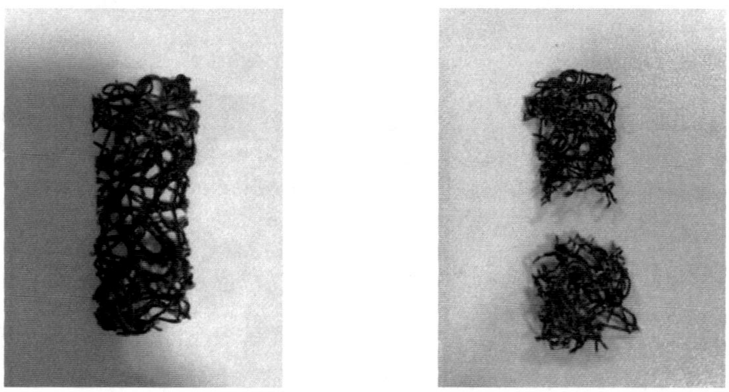

Figure 3. Sample before and after the test.

For each sample, the number of longitudinal filaments, which accounted for the tensile load, was calculated. The tests were carried out until all longitudinal filaments of the test sample were completely broken.

The results showed that the values of the maximum breaking load of the filaments of the considered samples of geomat are significantly different. This indicates a possible difference in the degree of microdamage of the filaments. Objective monitoring of microdamage of the filaments is carried out by computed X-ray tomography.

Using computed tomography, it becomes possible to build a three-dimensional model of the object under study, to obtain any of its sections and projections for further analysis. An YXLON Y.Cheetah high-resolution microfocus X-ray tomograph was used, equipped with an open type 20-160 kV x-ray tube. The characteristics of the tomograph are shown in Table 2.

Table 2. Characteristics of the tomograph.

Max.kV	Max.mA	Max.Tube Power	Max. Target	Min. FOD	Min. Focal Spot	Min. Detail Detectability

Output						
160	1.0	64W	10W	250 μm	< 2 μm	< 0,5 μm

To process images in real time and improve the contrast of images in shades of gray created in the X-ray system, the contrast setting is used.

Samples of geomat filaments were selected for research arbitrarily from the considered points of selection. Bunches are formed from the selected filaments. Using computed tomography, mathematically reconstructed 3D models of prepared filament samples are obtained. Figure 4 shows a 3D model of the test sample.

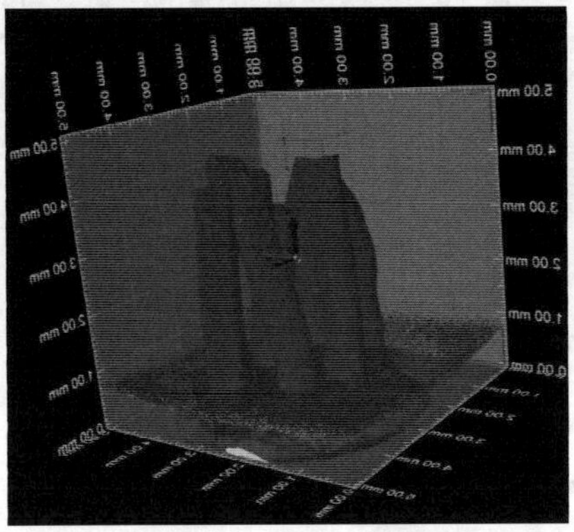

Figure 4. Reconstructed 3D model of the sample.

The presence/absence of micro/macrodamage of the geomat and its heterogeneities (inclusions differing in X-ray density in degree of grayness from the main material of the sample) was investigated. The dark areas of the resulting model correspond to the lowest X-ray density of the material – pores or defects in the limiting case, the brightest – to the most X-ray dense material. The brightest spots in the constructed images of the selected sections indicate the presence of X-ray dense inclusions in different sections of the samples of filaments. Intersections of microdamage of the material with different sections of the studied samples of threads appear in the form of dark lines.

Despite the fact that all samples of the material are now at the research stage, the results are shown for two samples taken at the sampling points P3U_Shd (the upper part of the slope, covered with steel mesh, in the shade) and P8D_Sun (the lower part of the slope, not covered with steel mesh, the sample was taken in the sun). The results of the tests are summarized in Table 3. Fiber strength correlates with the observed bulk density of microdamage.

Table 3. Test results.

Sample	Averaged maximum load per fiber (N)	Examples of characteristic longitudinal sections of individual filaments	Examples of characteristic cross sections of individual filaments	Comments
P3U_Shd	3.3±0.5	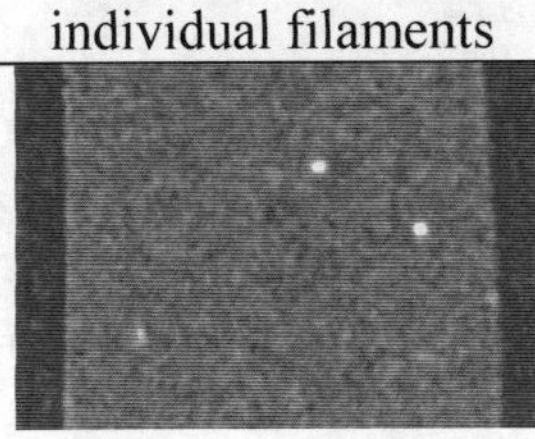	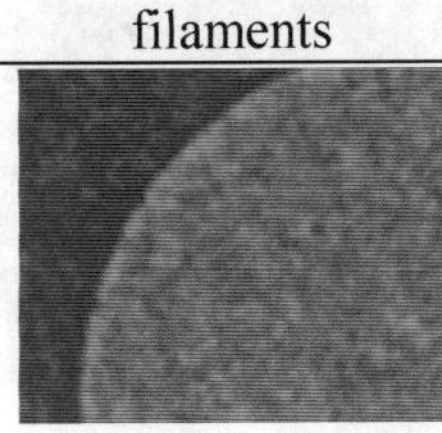	Minimum density of observed microdamage

P8D_Sun	1.5±0.5	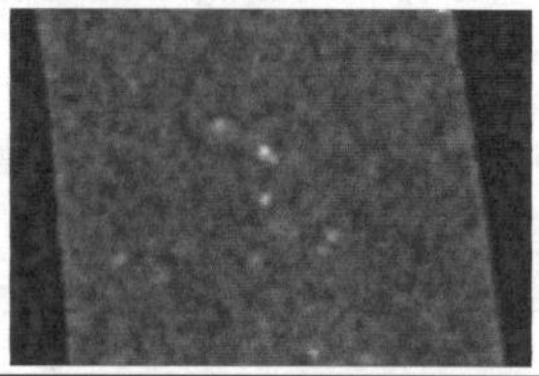	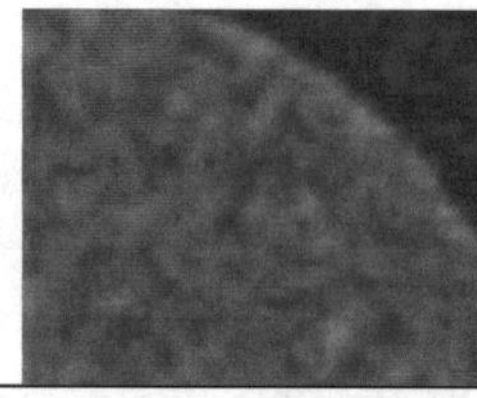	Maximum density of observed microdamage

4. Conclusion

Using an example of the complex of engineering measures for strengthening the slopes in the city of Svetlogorsk (Kaliningrad Region, Russia), an approach for assessing changes in the state of geosynthetic material used to hold landslide slopes, depending on actual operating conditions, is proposed. The approach is based on a rational choice of sampling points for geomat and objective control of structural and mechanical properties. The kinetics of changes in the properties of the geomat, which is used to stabilize sea slopes, and the determining factors of the studied changes in the characteristics of geomats of different types (corrosion) can be estimated from the results of regular monitoring. The results will form the basis for assessing the residual resource of the complex of engineering measures for strengthening the slopes, the selection of recommendations for its repair and optimization.

5. Acknowledgments

Sampling and analysis of operating conditions was supported by the project of the Russian Foundation for Basic Research No. 18-55-76002

References

[1] Esiukova E E, Chubarenko B V and Burnashov E M 2018 *Regional ecology* **3(53)** pp 7-20

[2] Esiukova E, Chubarenko B and Simon F-G 2018 *Proc. of 7th IEEE/OES Baltic Symp. "Clean and Safe Baltic Sea and Energy Security for the Baltic countries"* 12–15 June 2018, Klaipėda, Lithuania. IEEE Xplore Digital Library, pp 1-6

[3] Domnin D A, Burnashov E V, Esiukova E E and Karmanov K V 2019 *Aktual'nyye problemy nauk o Zemle: issledovaniya transgranichnykh regionov: Proc. IV Intern. Scient.-pract. Conf., dedic. to the 1000th anniv. of Brest*, Brest, Republic of Belarus, 12–14 september 2019 (Inst. of Nature Management of the National Academy of Sciences of Belarus, Brest State A.S. Pushkin Univ., Brest State Tech. Inst.; editorial: Karabanov A K, Bogdasarov M A, Volchek A A, Brest: Brest State Univ.) **2** pp 222-225

IOP Conf. Series: Materials Science and Engineering **911** (2020) 012005 doi:10.1088/1757-899X/911/1/012005

Analysis of early structure formation of anhydrite binders from secondary raw materials using microcalorimetry

L A Anikanova[1], O V Volkova[1], A I Kudyakov[1], V A Lotov[1] and D M Pervushina[1]

[1]Tomsk State University of Architecture and Building, 2, Solyanaya Sq., 634003, Tomsk, Russian Federation

E-mail: v.olga.nikitina@gmail.com

Abstract. This article presents results of a research on changes in energy parameters of fluorogypsum composition of normal consistency with introduction of various additives ($\mu\Delta n$) such as sodium sulfate and sodium sulfite. Activity of binder in interaction with water was defined as the sum of thermal effect of hydration and thermal effect of dispersion of fluorogypsum crystal lattice. In order to compare processes of heat release in the studied gypsum-containing systems, the article presents results of analysis for the following ones: fluorogypsum binder – water, fluorogypsum binder – sodium sulfate – water and fluorogypsum binder – sodium sulfite – water. Research has shown that processes of hydration during by development of alternating exothermic and endothermic effects in systems containing fluorogypsum binders.

1. Introduction

Taking into account significant needs of construction industry in material resources, Russia has developed a system of construction and technological utilization of industrial waste which includes ensuring environmental safety, as well as assessment, testing, classification and regional accounting of waste as a source of secondary raw materials for manufacturing products, component and technological forecasting; rationing, standardization of construction materials with maximal use of secondary raw materials [1, 2, 3]. Efficiency of secondary raw materials' usage depends on how much they are studied as potential raw materials for production of specific building materials.

In construction of buildings, alternative gypsum-containing secondary raw material – fluorogypsum – from chemical, nuclear and non-ferrous metallurgy enterprises can be used as finishing and wall material [4, 5]. The reason for slow introduction of fluorogypsum raw materials in production of construction materials is instability of its composition and properties, low activity and strength of the hardened stone, as well as insufficient knowledge of physical and chemical processes of its structure formation. Therefore, firstly, it is necessary to systematize and develop its classification by most important parameters and characteristics, and then distribute obtained data among the forming territorial-industrial complexes for production of construction materials [1, 3, 6].

Existing problems of utilization of fluorogypsum waste in production of construction materials are related to ambiguousness of available data on the mechanism of hydration and structure formation of binders based on it [7, 8]. As a rule, to intensify processes of hydration and hardening of anhydrite binders hardening activator additives are used. Identification of mechanisms of hydration and hardening processes that commence with introduction of additives is important for understanding processes of transition of the system to a colloidal state at first, and then to a crystalline one. For an efficient selection of external energy impact at an early stage of hydration process development, it is necessary to analyze alternations in energy parameters in the fluorogypsum binder – water system using known laws and tools. It is recommended to use Gibbs equation, the combined equation of the first and second laws of thermodynamics [9]:

$$\Delta G = \Delta H - T\Delta S = P\Delta V + \sigma\Delta s + \mu\Delta n + \varphi\Delta q - T\Delta S,$$

where **ΔG, ΔH, TΔS** – free energy, enthalpy and entropy factors of the fluorogypsum binder – water system accordingly.

Results of analysis of enthalpy and entropy factors' energy components help to select probable impacts on the fluorogypsum – water system managing complex physical and chemical processes of hydration and structure formation of fluorogypsum compositions. This article presents results of the research of energy parameter alternations in fluorogypsum composition of normal consistency after introduction of various chemical additives ($\mu\Delta n$). The energy released or absorbed during formation of new chemical compounds and changes in its phase composition are analyzed.

2. Research methodology and applied materials

Hydration of fluorogypsum binders is accompanied by heat release, which is spent on dispersion of particles over a long period of time and increase in entropy of the system. It is inefficient to use existing microcalorimeter designs to control system temperature changes, since they allow determining only the total amount of heat generated by the system. During the research, instrumental determination of heat release during early structure formation of fluorogypsum composite binders was performed using a differential microcalorimeter (DMC) [9-12] based on temperature difference between dry powder (reference sample) and fluorogypsum composition with a given water content located in two thermally insulated calorimetric cells. The DMC design allows using the principle of differential thermal analysis (DTA), which helps to control temperature difference between calorimetric cells over time. The resulting hydration heat is determined by temperature difference between the cells.

The following materials were used in the research: powdered fluorogypsum with a specific surface area of 400 m^2/kg neutralized with limestone flour. Mineralogical composition of fluorogypsum is represented by the following components: $CaCO_3$ – 5.0 %, CaF_2 – 6 %, $CaSO_4\,2H_2O$ – 10.0 %, $CaSO_4$ – 79.0 %. Sodium sulfate (GOST 6318-77) and sodium sulfite (GOST 5644-75) were used as hardening activators. During the research W/T=0.46 fluorogypsum binder with content of sodium sulfate and sodium sulfite in the amount of 2% of the weight of fluorogypsum was used.

3. Research results

Changes in temperature difference in the fluorogypsum binder – water system during hydration and hardening over the first 72 hours are given in Figures 1 – 3. The first intensive maximum of heat release in the fluorogypsum binder – water system (Figure 1) is due to a decrease in surface energy of the solid phase and release of hydration heat. Activity of binder in interaction with water is determined by the sum of thermal effect of hydration and thermal effect of dispersion of fluorogypsum crystal lattice with Q_d>Q_h. After contact of fluorogypsum binder without additives with water, wetting temperature is 1.7 ^{0}C, minor bursts of dispersion and crystallization are apparently associated with the presence of soluble form of calcium sulfate in fluorogypsum.

In this system, there is a cyclical repetition of processes of dissolution and crystallization of accumulated products every 24 hours during 3 days (Figure 1). Heat release kinetics of fluorogypsum binder with addition of sodium sulfate is also cyclical (Figure 2). After binder's contact with water, intense heat generation occurs during first minutes, and then the heat is spent on dispersion. Long second induction period of the binder with addition of sodium sulfate which lasts about 15 hours is apparently associated with dissolution of developed double salts. The processes of dissolution and crystallization are also cyclical. In all systems (Figures 1 – 3), after a temperature decrease in induction period, during which entropy factor prevails, temperature increases after a certain period of time due to release of crystallization heat.

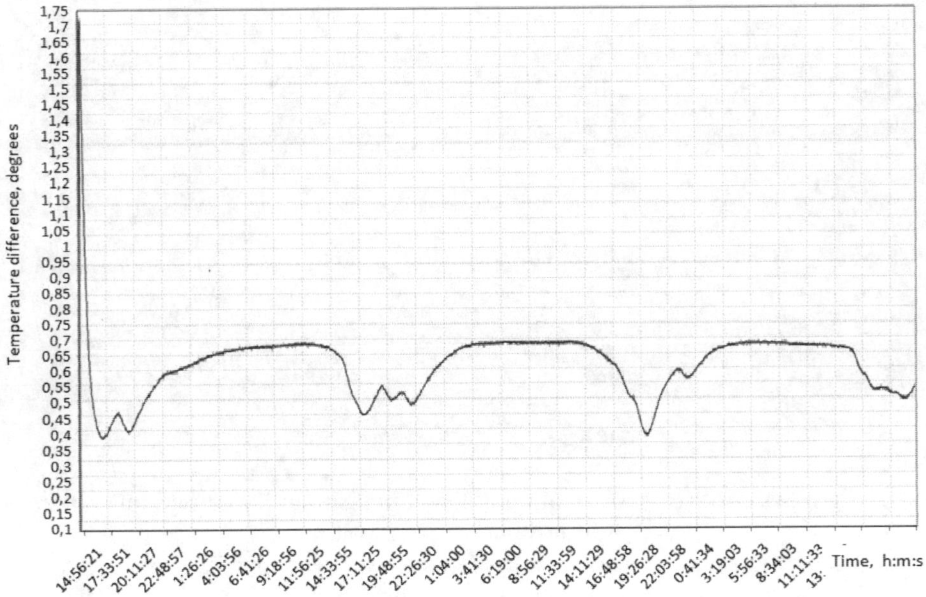

Figure 1. Kinetics of dissipation of the composition: fluorogypsum-water over 72 hours

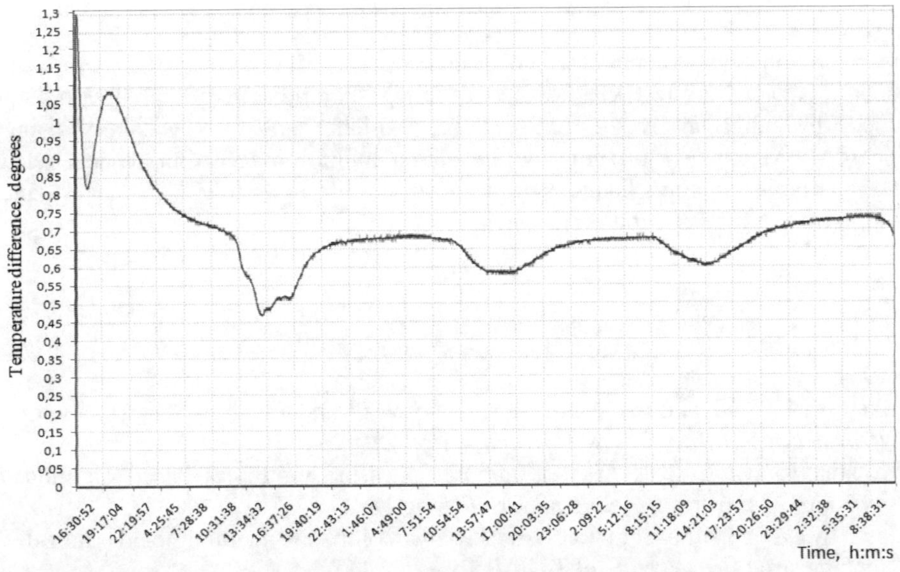

Figure 2. Kinetics of dissipation of the composition: fluorogypsum-sodium sulfate-water over 72 hours

Heat release kinetics of fluorogypsum binder with addition of sodium sulfite also has cyclical character, with dispersion process lasting about 3 hours and crystallization process lasting about 15 – 20 hours. With addition of sodium sulfite dissolution processes intensify and heat release processes are much faster (Figure 3). Formation of primary hydrated particles and their subsequent association due to coalescence is manifestation of binding properties of fluorogypsum binder.

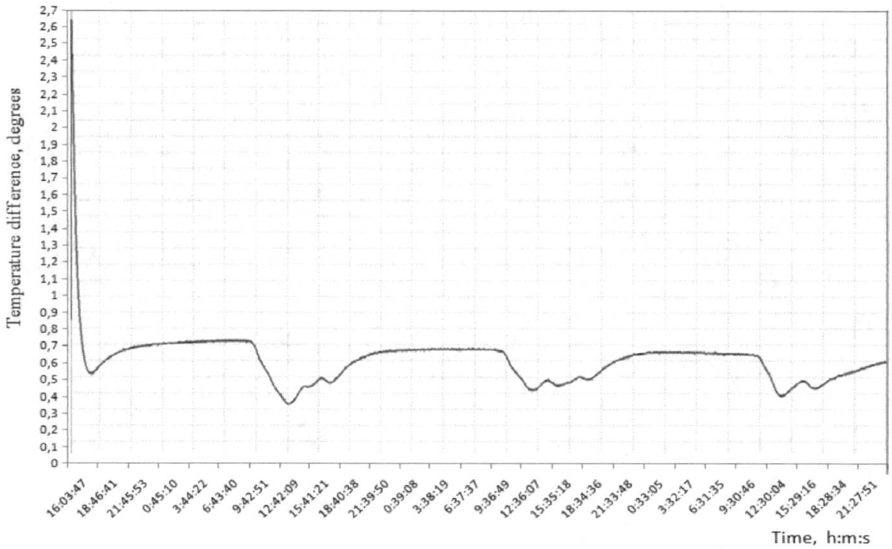

Figure 3. Kinetics of dissipation of the composition: fluorogypsum-sodium sulfite-water over 72 hours

Thus, it has been shown that processes of hydration during the first 72 hours are accompanied by development of alternating exothermic and endothermic effects, which is indicative of intermittence of processes developing in systems containing fluorogypsum binders.

4. Conclusion

Analyzing hydration and structure formation processes of fluorogypsum binders with additives of sodium salts, it is necessary to take into account energy components of enthalpy and entropy factors, accompanied by alternations in temperature, which can be measured using differential microcalorimetry. Mechanism of fluorohydrite binder hydration according to heat release kinetics is manifested in periodic topochemical colloidization of hydrating substances and subsequent crystallization of developed hydration products.

References

[1] Loganina V, Pyshkina I, Martiashin G 2017 Effect of supplement based on calcium hydrosilicates on the resistance of lime coatings *Magazine of civil Engineering* **4 (72)** pp 20-27

[2] Anikanova L, Kudyakov A, Volkova O 2015 Wall and finishing materials with fluorogypsum *Proceedings of Bratsk State University* **1** pp 230-234

[3] Petropavlovskaya V 2018 Use of mineral ultrafine modifiers based on industrial waste in gypsum composites *Building materials* **8** pp 18-24

[4] Kudyakov A, Anikanova L, Redlikh V 2012 Composite Bindung asid Fluoride Materials for fencing structures *Vestnik of TSUAB* **1** pp 106-111

[5] Anikanova L, Volkova O, Kudyakov A, Kurmangalieva A 2019 Activated composite fluorogypsum binder *Construction materials* **1-2** pp 36-42

[6] Anikanova L, Volkova O, Kudyakov A, Sarkisov Yu, Tolstov D 2016 Influence of solidification accelerators on structure formation of anhydrite-containing binders *IOP Conference Proceedings* **1** 070002

[7] Adewumi A, Ismail M, Ariffin M, Yusuf M, Maslehuddin M, Mohamed H 2019 Strength and microstructure of alkali-activated natural pozzolan and limestone powder mortar *Magazine of Civil Engineering* **92(8)** pp 36–47

[8] Kudyakov A, Anikanova L, Kopanitsa N, Gerasimov A 2001 Influence of grain composition and type of fillers on properties of building solutions *Building materials* **11** pp 28-29

[9] Jassim Muhsin, Abdul Qader Nihad, Arshad Nadhom Ali 2019 Effect of adding industrial wastes on the mechanical properties of gypsum *International Journal of Science and Research (IJSR)* **8** pp 2123-2125
[10] Lotov V 2018 On interaction of cement particles with water or a mechanism of the cement hydration and hardening processes *News of Tomsk Polytechnic University. Engineering of georesources* **1** pp 99-110
[11] Usherov-Marshak A, Sopov V 2009 Isometric calorimetry: a standard method for studying cement hydration kinetics *Cement and its application* **5** pp 106-107
[12] Moser B 2010 Progress in building materials analysis *ZKG Zement Kalk Gips International* **2** pp 63-72

Experimental studies of the evolution of structural and mechanical characteristics of cement stone in the process of hydration

M A Dmitrieva[1], A V Sharanova[1] and V N Leitsin[1]

[1]Immanuel Kant Baltic Federal University, 14 Nevskiy St, Kaliningrad, 236016, Russian Federation

E-mail: dmitrieva_m@inbox.ru

Abstract. An experimental approach is proposed to study the evolution of structural and mechanical characteristics of cement stone in the process of hydration. This approach is based on the use of computed tomography methods and the determination of the strength at different times of the setting of concrete compositions. It is shown that modifying additives of calcium formate and redispersible polymer powder provide an increase in strength at the initial setting time. In this case, the highest strength value 30 minutes after mixing is achieved for a composition with 2% content of redispersible polymer powder by weight of cement. The highest density is demonstrated by compositions with additives of redispersible polymer powder. A developed experimental approach, combining a non-destructive computed tomography method and a direct assessment of the compressive strength of samples of multicomponent concrete of various compositions, makes it possible to effectively evaluate the effect of modifying additives on the properties of modified concrete.

1. Introduction

The existing standard methods for studying the internal structure of concrete are based on the use of microscopy. By layer-by-layer analysis of sample sections, it is possible to determine the location and nature of the pores, as well as the quality of the formed spatial bonds in the cement stone. However, this technique has a number of significant drawbacks, which include not only the laboriousness in performing operations for preparing sections of samples, leading to their destruction, which makes it impossible to quickly study the evolution of the structure of individual samples at different stages of hydration. Classical methods for controlling the structure of concrete cannot be used in studies of changes in the structure of materials at the early periods of setting, and obtaining a clear and complete picture of the location of pores, voids and cracks using these methods is impossible [1]. The development of non-destructive control methods is an urgent task of studying the process of structure formation of concrete. The use of X-ray computed tomography for these purposes allows examining the complete structure of the material without subjecting the sample to mechanical destruction.

The use of various instruments based on computer X-ray tomography in modern materials science is becoming one of the main methods for studying the structure of inhomogeneous materials due to a large number of advantages. This method makes it possible, for example, to completely eliminate mechanical stress on the samples. The software helps not only to visually characterize the location of individual phases and pores in the samples, but also to distribute their dimensions [1, 2]. A similar result of studying the structure cannot be provided by any of the methods used for controlling the structure of heterogeneous materials. For example, the analysis of pore sizes in multicomponent concrete samples modified with nanoparticles, as well as the nature of their distribution in the material, made it possible to conclude that the additives also have a positive effect on the microstructure of cement stone, making it denser [3]. Experience shows that from the point of view of research using computed tomography methods, in addition to analyzing the effect of modifications on the properties of a composition, a set of survey parameters should also be taken into account [2, 4].

In addition to studying the structure formation of concrete at the stage of working out the mixture in laboratory conditions, computed tomography is also actively used by researchers in the analysis of concrete samples that have been subjected to operation. By means of computed tomography, it is possible to determine the nature of destruction throughout the volume of the material, the magnitude of crack opening. Researchers from Japan conducted a series of experiments related to a detailed study of the features of visualization of images from samples without any damage to the obvious presence of cracks in them as a result of operation [5]. In addition to the visible nature of the propagation of changes in the volume of the material, it was also revealed how various kinds of kinks relate to the value of the absorption coefficient of X-ray radiation.

2. Materials and methods

The following materials were used to prepare the concrete mix: 1) Portland cement (PC) Eurocem super 500, CEM I 42.5H, Petersburg Cement; 2) Sand: unwashed construction, fractions: 1.25 mm and 0.315 mm; origin location: Kaliningrad region; 3) Silica fume (SF): wastes of the metallurgical industry, country of origin: Poland; 4) Highly active metakaolin (HAM): white, country of origin: Russia, Chelyabinsk region, Synergo; 5) Polypropylene fiber: length of 12 mm, thickness of 35 μm, country of origin: Russia; 6) STACHEMENT 1267 hyperplasticizer based on polycarboxylates for the production of ready-mixed concrete: country of origin: Poland; 7) Calcium Formate (CF); 8) Redispersible polymer powder (RPP). Material consumption is shown in Table 1.

Table 1. Component consumption per 1 m^3.

Components	Concrete mix number			
	1	2	3	4
Cement (kg)	545.3	545.3	545.3	545.3
Sand (kg)	1168	1168	1168	1168
Water (l)	416.6	381.9	381.9	381.9
Silica fume (kg)	156	156	156	156
Metakaolin (kg)	78	78	78	78
Calcium formate (kg)	-	10.4	-	-
Redispersible polymer powder (kg)	-	-	10.4	5.2
Fiber (kg)	1.2	1.2	1.2	1.2
St. 1267 (l)	11.7	11.7	11.7	11.7

In the course of the study, samples of concrete mixtures of a cubic shape with a size of 20x20x20 mm were made, for each mixture the setting times and compressive strength were determined at the early stages of hardening [6]. The dry components were mixed by hand until smooth, then polypropylene fiber was added, pre-divided into individual fibers. Next, the plasticizer was diluted in the required volume of water, after which the dry mixture was mixed. The concrete mixture was placed in molds treated with a lubricant and carefully compacted to extract excess air, which, in particular, could get into the mixture when it was stirred. After the moment of mixing with water, the countdown began until the moment of strength tests. The samples were tested for compressive strength at the ages of 30, 45, 60, 75 and 90 minutes. To carry out strength tests, a universal testing machine INSTRON ElectroPuls E1000 was used - a benchtop machine with Bluehill software and a wide range of speeds for simulating real conditions.

IOP Conf. Series: Materials Science and Engineering **911** (2020) 012006 doi:10.1088/1757-899X/911/1/012006

The internal structure and the nature of the destruction of the samples were analyzed using X-ray computed tomography on a YXLON Y. Cheetah instrument.

The shooting process consists in the selection of parameters for the location of the ray tube and the detector relative to each other. It is this setting that significantly affects the level of contrast and color intensity in future images. The shooting area can be further corrected in relation to the location of the sample in the tomograph camera. After the completion of the sample survey, the results are loaded into a graphics station, using the software "Volume Graphics Studio" of which it is possible to build a 3D model of the material and analyze specific structural elements throughout the entire volume of the material under study, as well as additionally adjust the image parameters.

3. Results

The results of testing the compressive strength of a series of concrete specimens of various compositions in the early stages of hardening are presented in Table 2. All reported results have an error within ± 10%.

Table 2. The strength of the compositions at the initial stages of hardening.

Concrete mix number	Strength value, MPa				
	30 min	45 min	60 min	75 min	90 min
1	0.06	0.08	0.10	0.11	0.13
2	0.08	0.10	0.13	0.17	0.16
3	0.12	0.17	0.17	0.17	0.20
4	0.09	0.13	0.14	0.17	0.18

Various kinetics of the strength gain of the samples under study are observed. The kinetics of setting concrete of composition No. 3, which has an inflection point, is of interest for further research.

The typical structure of the samples for the 90th minute of setting is shown in Figure 1.

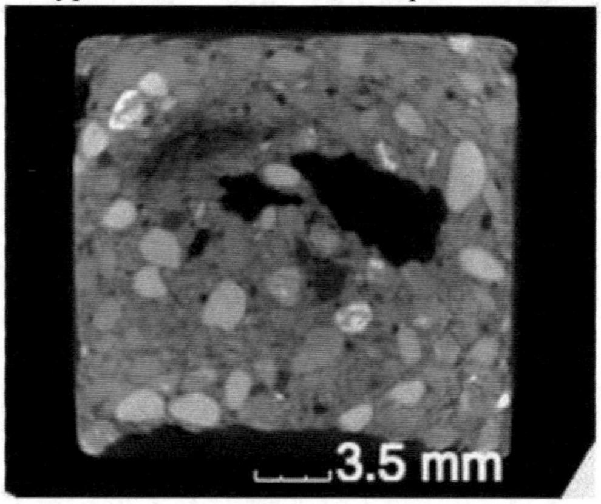

Mix No. 1

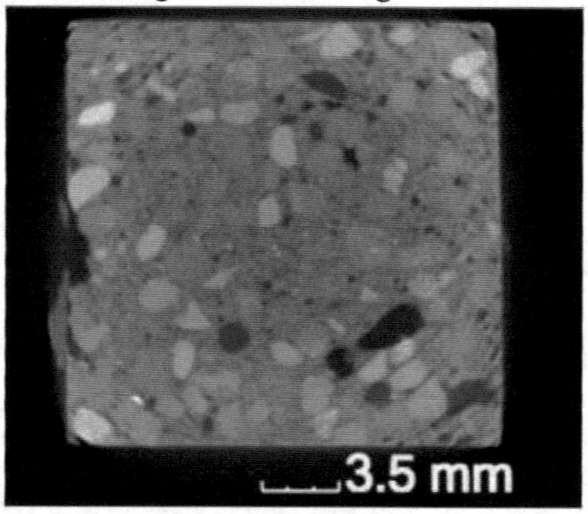

Mix No. 2

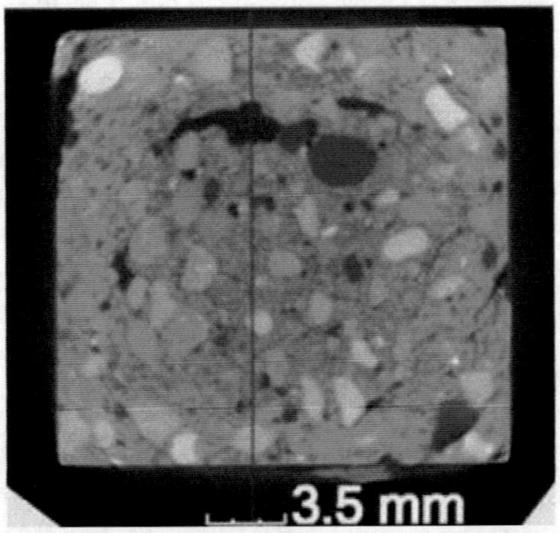

Mix No. 3

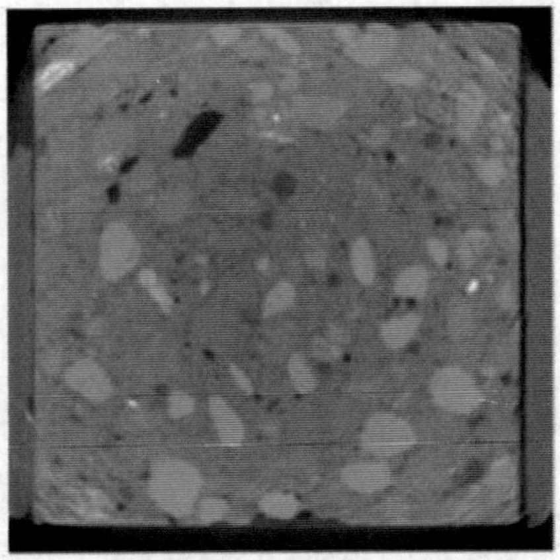

Mix No. 4

Figure 1. Structures of samples at 90 minutes of setting

In the case of concrete, the following component categories can be distinguished with respect to the grayscale in the resulting images. Light gray represents the cement matrix and dark gray represents the aggregate. Based on the level of X-ray absorption by the material, pores, cracks and voids are black. For any section of the samples represented by the obtained 3D model, individual components of the cement stone, cracks and pores are clearly observed.

For a series of images of ordinary sections of samples of each of the compositions, the following indicators were determined: the number of voids, their average size, their percentage of the total cross-sectional area of the sample, as well as the total area and the minimum value of the pore diameter from the obtained sample. The averaged indicators of both porosity and other values in the series for each of the compositions are presented in Table 3.

Table 3. Average structure characteristics of samples.

Structure characteristic	Concrete mix number			
	1	2	3	4
Porosity (%)	6.37	4.65	3.45	2.84
Number of voids (pcs)	407	369	277	372
Occupied area (mm^2)	24.8	18.98	14.34	11.7
Average size of voids (mm^2)	0.06	0.05	0.05	0.03
Minimum diameter (mm)	0.15	0.15	0.14	0.12

4. Discussion

All compositions modified with RPP and CF additives have a general tendency to increase the strength at the initial setting time compared to the control sample. The highest strength value 30 minutes after mixing is achieved for composition No. 3 with 2% of RPP.

For a more accurate interpretation of the results of the influence of additives on the initial composition of the studied composition, it is undoubtedly necessary to consider the physical and mechanical properties of these samples after they reach the design strength at the age of 28 days.

For composition No. 4 with 1% RPP content, the characteristics related to the dimensions of voids and porosity has the lowest value among all the samples presented. Composition No. 1 (without the addition of RPP and CF) has the worst performance among all the studied compositions. However, the average void size is only slightly higher than that for compositions using calcium formate and 2% RPP. Among the three compositions with chemical additives, the one with 1% RPP content has the best performance. Despite the fact that the total number of recognized particles in the fourth composition is greater than in the third, the voids occupy a smaller area and have a smaller size, which indicates a lower probability of the presence of large pores in the material volume.

Thus, the experimental approach to study the evolution of the structural and mechanical characteristics of cement stone during hydration, based on the use of computed tomography methods and determining the strength at different times of setting of concrete compositions, makes it possible to effectively evaluate the effect of modifying additives on the properties of multicomponent concretes.

Acknowledgment

All presented experimental results were obtained with the active participation of graduates of the bachelor's degree in Civil Engineering of Immanuel Kant Baltic Federal University: Kogay V.V., Lenkova D.A., Panfilova A.D., Shinyaeva M.V.

The reported study was funded by RFBR and Kaliningrad region, project number 19-48-390001 "Multicomponent nanomodifiers of concrete mixes".

References

[1] Osipov S P, Prishchepa I A, Kudyakov A I, Batranin A V and Osipov O S 2018 Computed tomography of foam concrete *Systems. Methods. Technology* **2** p 146-152

[2] Osipov S P, Klimenov V A, Batranin A V, Stein A M and Prischepa I A 2015 Digital radiography and x-ray computerized tomography in building construction and construction materials science *Vestnik of Tomsk state university of architecture and building* **6** p 116-127

[3] Rajczakowska M, Stefaniuk D and Lydzba D 2015 Microstructure characterization by means of X-Ray Micro-CT and nanoindentation measurements *Studia Geotechnica et Mechanica* **37(1)** p 75-84

[4] Peng R, Yang Y, Ju Y, Mao L and Yang Y 2011 Computation of fractal dimension of rock pores based on gray CT images *Chinese Science Bulletin. Mechanical Engineering* **56 (31)** 3346

[5] Suzuki T, Aoki M and Ohtsu M 2010 Damage identification of cracked concrete by X-Ray computed tomography method *Fracture Mechanics of Concrete and Concrete Structures* p 1143-1148

[6] Dmitrieva M, Leitsin V, Sharanova A and Tovpinets A 2018 Complex approach to the study of modern concrete characteristics *MATEC Web of Conf.* **251** 01024

IOP Conf. Series: Materials Science and Engineering **911** (2020) 012007 doi:10.1088/1757-899X/911/1/012007

Optical properties of sol-gel coatings containing silver nanoparticles

S N Stepin[1], T A Xolmurodov[1], O O Mirzaev[1] and V E Katnov[1]

[1]Kazan National Research Technological University, Kazan, Russian Federation

E-mail: tima.xolmuradov@mail.ru

Abstract. The technology of receiving the antireflective coating increasing efficiency of use of sunlight in various service conditions is described. Suspensions of nanodimensional silicon dioxide with tetraethoxysilane as a precursor were synthesized to solve the assigned task. The received suspensions were applied by a dipping method on glass samples followed by the coating formation on a surface of the substrates at various temperatures. As a result of a cure of the received samples under natural conditions transparent film of about 500 nanometers thick having very low strength properties on the surface of glass were formed. Heat treatment led to the improvement of strength of coatings in process with increase of temperature up to 600 °C. Exceeding of this temperature resulted in cracking of coatings. In the course of heat treatment higher than 300 °C formation of a coating was accompanied by the thermolysis of silver nitrate that was introduced into silicon dioxide suspension with a formation of silver particles which size was in the nanoscale distance limits according to the results of Auger electron spectroscopy. Presence of evenly distributed silver nanoparticles provided an antireflective effect to coatings which at visual observation found expression in the form of the interference phenomenon. The obtained data confirmed prospects of use of nanosilver as a component of coatings on the basis of silicon dioxide in order to confer the antireflective properties.

1. Introduction

All currently widely used optical systems that allow the transmission of light energy and images are based on the effects of reflection and refraction. Reflection increase means the reduction of light transmittance through a transparent medium. For example, about 8.6% of the incident light at an angle of 0° is reflected from the two interfaces of glass and air, when the refractive index of silicate glass n = 1.51.

Improving the efficiency of optical devices is an urgent issue [1-3]. The antireflection coating application (or anti-glare coating) is used in areas where light passes through an optical element and it is necessary to reduce the loss of intensity or eliminate reflection (lenses, solar cells, etc.). The most well-known solution to the problem of reducing reflection and increasing light transmittance is the use of low-reflection thin-film interference coatings, the main limitation of which is the impossibility of their application on the large-size optical components. In addition, it is necessary to create multilayer coatings of this type in order to achieve the antireflective effect in a wide range of wavelengths.

Therefore, other ways to reduce reflection and thereby increase the light transmittance of optical objects are searched for. The common cases are eyeglass lenses and camera lenses.

In this study, the low-reflection properties of sol-gel coatings formed on the basis of silicon dioxide and the influence of dispersed nanoscale silver on their optical properties were investigated. The possibility of using nanosilver in low-reflective coatings was shown in the studies, where coatings based on polymethyl methacrylate were used as an example [4, 5].

2. Experimental

At the first stage of the study, aqueous nanosuspensions of silicon dioxide were synthesized, which were the precursor for obtaining the investigated coatings. Synthesis of SiO_2 sols was carried out according to the known method, including the hydrolysis of tetraethoxysilane (TEOS) and polycondensation according to the reaction equations:

$$\equiv Si{-}O\,C_2H_5 + H_2O = \equiv Si{-}OH + C_2H_5OH;$$
$$\equiv Si{-}O\,C_2H_5 + HO{-}Si\equiv\; = \;\equiv Si{-}O{-}Si\equiv + C_2H_5OH;$$
$$\equiv Si{-}OH + HO{-}Si\equiv\; = \;\equiv Si{-}O{-}Si\equiv + H_2O,$$

The process of hydrolysis of TEOS occurs most quickly and completely when an appropriate catalyst is used. To obtain silicon dioxide of various structures, two types of synthesis catalyst were used, viz. hydrochloric acid as an acid catalyst and ammonia as alkaline one.

The sols were synthesized by successively mixing a solvent (ethanol), distilled water, a catalyst, and introducing TEOS with continuous stirring. The resulting mixture was stirred for 24 hours until a homogeneous sol was obtained.

During the preliminary experimental studies, the following molar ratios of the components of the reaction mass were selected as optimal ones, TEOS / water / ethanol / hydrochloric acid = 0.25 / 0.95 / 8.00 / 0.10 and TEOS / water / ethanol / ammonia = 0.25 / 1.15 / 8.00 / 0.10.

The study of the dispersion composition of the obtained sols was carried out by the method of dynamic light scattering (DLS) using a Brookhaven 90Plus nanoparticle size analyzer.

3. Results and discussion

Figure 1 shows the lognormal distribution of particles by volume in the SiO_2 sol (a) and the bimodal distribution of particles per volume unit according to the data of the DLS (b), which is characterized by the presence of two main fractions.

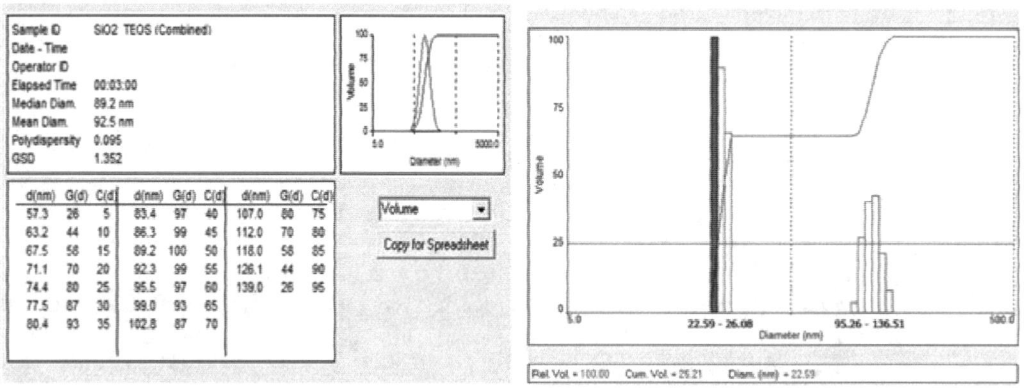

a b

Figure 1. Particle distribution by volume in SiO_2 sol obtained in the presence of an alkaline catalyst.

Figure 1b shows that the suspension contains mainly (70 vol.%) nanoparticles, the size of which is between 22.5–26.0 nm. However, particles of a larger size are also present in the suspension (30 vol.%), they correspond to the second maximum on a bimodal distribution curve with particle sizes of 95.0 - 136.5 nm.

Figure 2 shows the unimodal distribution of particles by volume in the SiO_2 sol obtained in the presence of an acid catalyst (a) and the bimodal distribution according to the data of the DLS (b), which is characterized by the presence of two fractions.

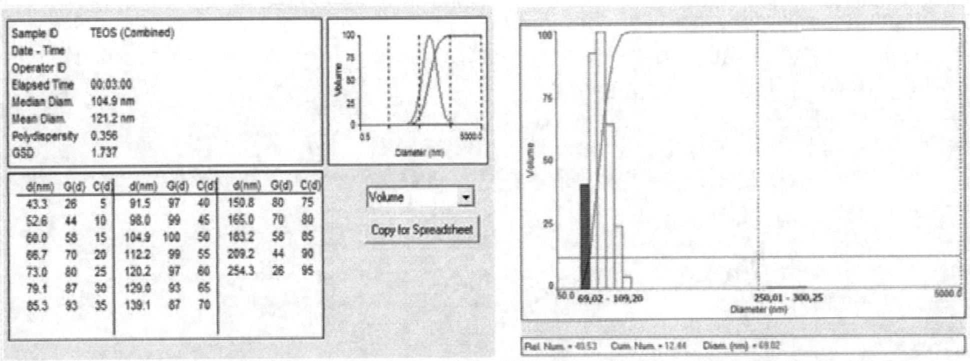

a b

Figure 2. Particle distribution by volume in SiO_2 sol obtained from TEOS in the presence of an acid catalyst.

According to Figure 2b, it can be said that the sol contains mostly (95 vol.%) isolated nanoparticles, the size of which is between of 69.00 - 109.00 nm. However, an insignificant proportion (5 vol.%) of aggregates is present in the suspension, it corresponds to the second maximum on the bimodal distribution curve with particle sizes of 205.00 - 300.00 nm.

Coatings based on the obtained suspensions were applied onto the stage glass by a dipping method, thus varying the rate of substrate extraction from the bath with the precursor, using RM-50 tension testing machine. The coating thickness was determined by using probe microscopy.

The study of the optical properties of the coatings was carried out by recording curves of the light reflection and light transmittance of the obtained coatings and painted glasses in the 400-700 nm wavelength range using X-Rite Color and Proscan MC 122 spectrophotometers, respectively. Pure glass similar to the one on which the coatings were applied was used as a reference standard when measuring optical characteristics.

Figure 3 shows the light transmittance spectra of coatings based on the sol prepared with an alkaline catalyst.

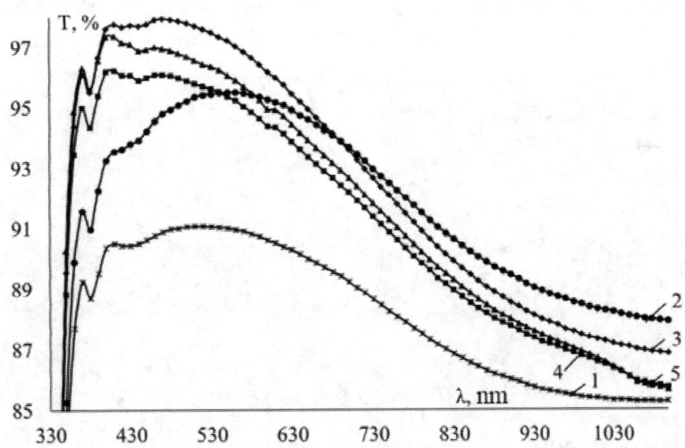

1 - uncoated glass; 2 –3%; 2–2%; 4 - 1%; 5 - 5%.

Figure 3. Transmittance spectra of glasses with SiO_2 coatings, obtained by dipping in sols with different concentrations (alkaline catalyst).

The obtained data allow us to conclude that coatings based on the SiO_2 sol synthesized in the presence of an alkaline catalyst increase the light transmittance of glass (an antireflective effect is observed) in the wavelength range of 350-1100 nm. The largest increase in optical transmittance is recorded for samples with a 2% concentration of SiO_2 in the resulting sol (curve 3), with a maximum of about 97.94% at a wavelength of 450–500 nm (Figure 3).

Figure 4 shows the light transmittance spectra of coatings based on a sol obtained from TEOS in the presence of an acid catalyst.

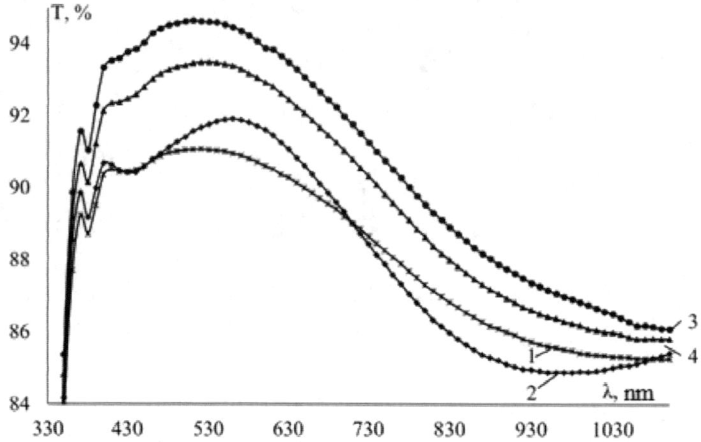

1 - uncoated glass; 2–5%; 3–3%; 4–1%.

Figure 4. Transmittance spectra of glasses with SiO_2 coatings, obtained by dipping in sols with different concentrations (acid catalyst).

Coatings based on the SiO_2 sol obtained in the presence of an acid catalyst also increase the light transmittance of glass (the antireflective effect) in the wavelength range of 350-1100 nm. The greatest increase in optical transmittance is observed in the 3% sol of SiO_2 (curve 3), with a maximum of about 94.7% at a wavelength of 450-500 nm.

Further studies used a silica sol obtained in the presence of an alkaline catalyst, which had a large low-reflection effect, according to the research.

An Olympus LEXT OLS4000 3D Laser Measuring Microscope was used to evaluate the external appearance of the obtained coatings. Micrographs of coatings of sols with different contents of silicon dioxide obtained in the presence of an alkaline catalyst are shown in Figure 5 and 6.

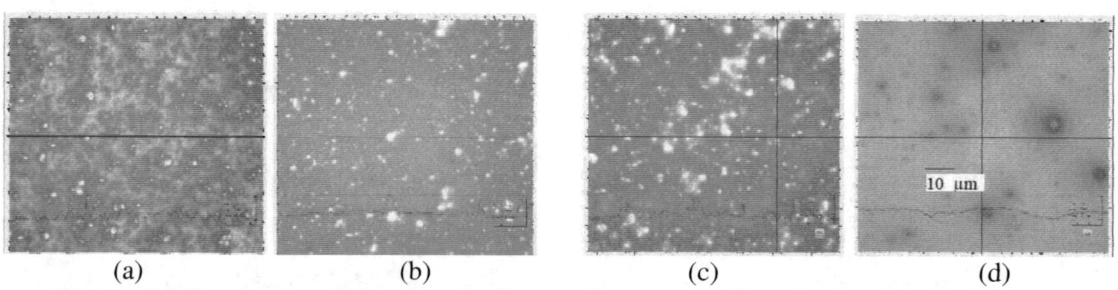

(a) (b) (c) (d)

Figure 5. Micrographs of glasses with SiO_2 coatings obtained by dipping in sols of various concentrations (alkaline catalyst) (% by mass): (a) – 1%, (b) –2%, (c) –3%, (d) – 5%.

According to the data presented in Figure 5, we can conclude that the obtained coatings have a certain texture, which varies depending on the content of silicon dioxide in the precursor.

A glass with a coating of 2% silica sol prepared in the presence of an alkaline catalyst was taken as a reference sample (Figure 3, curve 3).

Figure 6 show the transmittance and reflection spectra of glasses coated by silica with different silver contents regarding the reference sample. Visual observations from different view angles permitted to establish that the coatings had an interference effect.

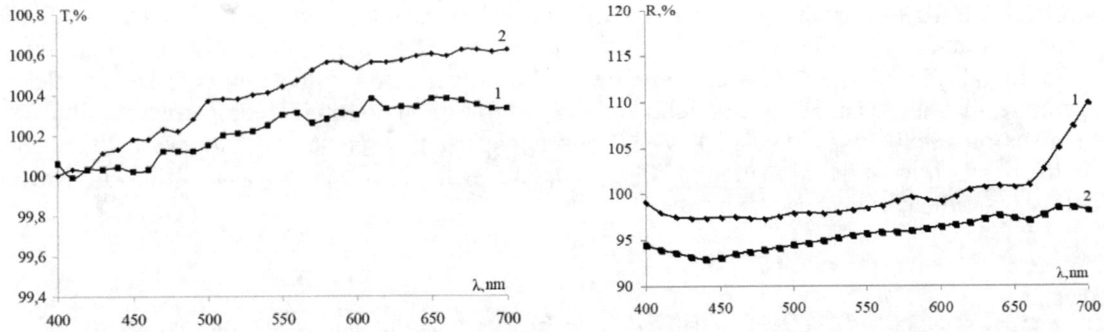

(a) (b)

Figure 6. Transmittance (a) and reflectance (b) spectra of coated glass samples obtained at the ratio of silica sol/silver being 1–10: 3 and 2–10: 1, regarding the reference sample.

As it can be seen from Figure 6a, the introduction of silver increases the transmittance and reduces reflection, but the ratio silica/silver being 10: 1 has a more favorable effect on this indicator. This ratio was used later on.

The insignificant antireflection effect after the introduction of nanosilver, suggests that calcination results in leveling of the surface of the coatings, in other words, in a decrease of the antireflection effect from its texture, as compared with the previously obtained data for polymethylmethacrylate coatings. Therefore, the result of the experiments is the net effect of the summation of effects from the introducing nanosilver into the composition of coatings and changing the texture of their surface.

The comparison results of reflectivity and light transmittance of coatings obtained at different temperatures are an indirect proof, they are presented in Figure 7. This comparison allows us to conclude that when the calcination temperature increases, the reflectivity of the coatings increases, and the light transmittance decreases, which is most likely due to the formation of their texture.

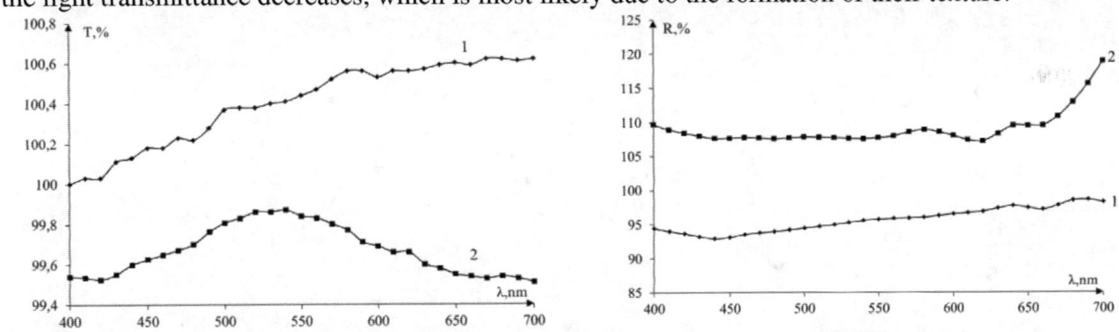

(a) (b)

Figure 7. Transmittance (a) and reflectance (b) spectra of glass samples with coatings obtained at the calcination temperature 1-500 °C and 2-600 °C.

4. Conclusion

Determination of a particulate composition of the synthesized silicon dioxide sols by dynamic light scattering showed that in the presence of the acid catalyst larger particles are formed. Their existence has negative effect on aggregative and sedimentation stability of the studied colloidal systems. In turn, the analysis of transmittance spectra of glass plates coated with received sols at different concentration of silicon dioxide allowed to make a conclusion about higher antireflective effect at coatings on the basis of the sol received with use of an alkaline catalysis. As a result of visual inspection of coatings by means of an optical microscope the specific texture of their surface which depends on the content of silicon dioxide nanoparticles was found. It is established that excess of a certain concentration of

IOP Conf. Series: Materials Science and Engineering **911** (2020) 012007 doi:10.1088/1757-899X/911/1/012007

nanoparticles leads to a cracking of coatings as a result of reaching of the critical value of internal tensions. Introduction to the structure of silver nanoparticles to the coatings formed at the room temperature allows to improve optical properties, however upon calcinating of the samples at temperatures higher than 500 °C a tendency to deterioration in optical properties is observed. Generalization of experimental data allows to draw a conclusion on prospects of the research aimed to study the regularities of the formation of sol-gel coatings with silver nanoparticles for use in the field of optical instrumentation and solar power.

References

[1] Ghosh S S, Das S, Sil A and Biswas P K 2012 Characterization of individual layers of an optical design based multilayered antireflection coating developed by sol–gel method *Journal of Sol-Gel Science and Technology* **64** pp 534–42

[2] Rahman A, Ashraf A, Xin H, Tong X, Sutter P, Eisaman M D and Black C T 2015 Sub-50-nm self-assembled nanotextures for enhanced broadband antireflection in silicon solar cells *Nature Communications* **6**

[3] Groep J van de, van de Groep J, Spinelli P and Polman A 2015 Single-Step Soft-Imprinted Large-Area Nanopatterned Antireflection Coating *Nano Letters* **15** pp 4223–8

[4] Gadomsky O N, Stepin S N, Katnov V E and Zubkov E G 2013 Nanostructured Composite Layers With Quasi-Zero Refractive Index *Journal of Applied Spectroscopy* **80** pp 726–30

[5] Gadomsky O N, Altunin K K, Stepin S N, Katnov V E, Rusin A A and Pereskokov E A 2014 Near-field effect in composite nanomaterials with a quasi-zero refractive index *Optics Communications* **315** pp 286–94

Insitu determination of continuous evolution in space and during the strain fields on steel plate with the stress concentrator

V A Vlasov[1], A S Plyaskin[2], A A Klopotov[3], A I Potekayev[4], A M Ustinov[5], Yu A Abzayev[6], A Kurgan[7], V E Bunkov[5, 8] and N A Matskevich[5]

[1]Doctor of Physical and Mathematical Sciences, Professor, Rector of the University ,Tomsk State University of Architecture and Building, Tomsk, 2 Solyanaya Square, 634003, Russian Federation
[2]Ph.D. tech. sciences, associate professor, head of the department "Metal and wooden structures", Tomsk State University of Architecture and Building, Tomsk, 2 Solyanaya Square, 634003, Russian Federation
[3]Doctor of Physical and Mathematical Sciences, Professor, Tomsk State University of Architecture and Building, Tomsk, 2 Solyanaya Square, 634003, Russian Federation
[4]Doctor of Physics and Mathematics, Professor, National Research Tomsk State Universty, Tomsk, 36 Lenin Avenue, 634050, Russian Federation
[5]Tomsk State University of Architecture and Building, Tomsk, 2 Solyanaya Square, 634003, Russian Federation
[6]Doctor of Physical and Mathematical Sciences, Professor, Tomsk State University of Architecture and Building, Tomsk, 2 Solyanaya Square, 634003, Russian Federation
[7]National Research Tomsk State University, Tomsk, 36 Lenin Avenue, 634050, Russian Federation

E-mail: rector@tsuab.ru, Plyaskinandrei@mail.ru, klopotovaa@tsuab.ru, potekaev@spti.tsu.ru, ustinov_a_m@tsuab.ru, abzaev2010@yandex.ru, kirill_k2.777@mail.ru, Viktorbunkov@inbox.ru, nina-dukvina@yandex.ru

Abstract. The article represents the results of the experimental research of the evolution of strain fields of a steel plate with a stress concentrator by the correlation method of digital images. The obtained results reflect elaborately the evolution of the distribution of isofields towards the strains on the surface of the steel plate and the behavior of their distributions, which corresponds to each stage of the stress-strain curve under uniaxial tension.

1. Introduction

For the reliable stress assessment of substructures and structural components as well as for the predictive modeling of their residual lifetime, the reliable experimental data on the distribution of stress-strain states are required in the field of stress concentrators. The bearing capacity of structural elements is determined

IOP Conf. Series: Materials Science and Engineering **911** (2020) 012008 doi:10.1088/1757-899X/911/1/012008

by the distribution of stresses and strength conditions in the places with the maximum stress value, since the destruction occurs particularly there [1].

The development of adequate ideas and making the correct calculation models used in the design of reliable structures are impossible without the detailed experimental data on the stress-strain states of stress concentrator fields under conditions of elasticity, ductility, creeping and damage. Recently, such data have been successfully obtained using insitu studies of displacement and strain fields with the using the Vic-3DCorrelatedSolutions digital optical system [2-7]. The developed ideas are successfully used in modeling the processes of strain effects on the structural elements with stress concentrators, which have been made of various metallic and non-metallic materials.

The aim of the current work is to study the continuous insitu evolution in space and during the distribution of stress-strain states of structural elements in the field of stress concentrators under the tension of plates made of 09G2S structural steel.

2. Material and methodology

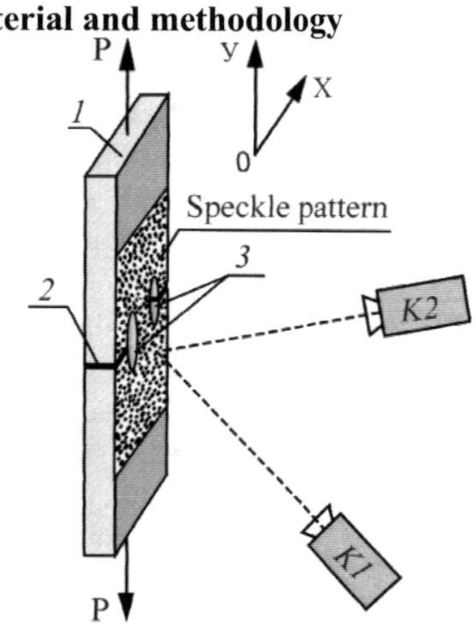

Figure 1. The scheme for the image registration from a speckle pattern on a plate with the symmetrical undercuts during the tensile strain is presented: 1–steel plates; 2 –undercut; 3 – holes; K1 and K2 –digital cameras; P - applied load.

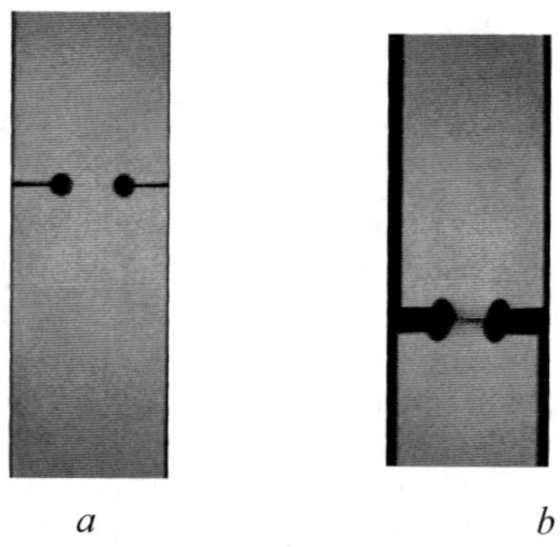

Figure 2. The picture of the central part of the sample is shown before testing (*a*) with a stress concentrator and after the testing (*b*).

The prototype models of the plates were made of 09G2S structural steel with a ferritic-bainitic-martensitic structure that combines a complex of mechanical properties. The choice of steel grade is justified by the wide practical application in the capacities of the petrochemical industry, operated in Siberia and the Far North [8]. The chemical composition of steel is shown in the table 1.

For testing of 09G2S steel, the prototype models were made with the length of 270 mm, the width of 40 mm, and the thickness of 6 and 8 mm. The stress concentrator was formed by two undercuts, located symmetrically towards the axis of the plate, to the depth of 15 mm each. The radius of wryness of the undercuts is formed by the holes with a diameter of 6 mm. The neck width of the residual cross-section formed by the undercuts was 10 mm (figure 2).

Table 1. The chemical composition of 09G2S steel (weight %).

IOP Conf. Series: Materials Science and Engineering **911** (2020) 012008 doi:10.1088/1757-899X/911/1/012008

Chemical element	N	Si	Mn	Cu	As	Fe
Density, weight%)	0.008	0,5–0,8	1,3–1,7	0.3	0.08	96-97
Chemical element	Ni	S	C	P	Cr	-
Density, weight%)	0.3	0.04	0.12	0.035	0.3	

The tensile testing of the samples was carried out on an "INSTRON 3386" testing machine with a constant strain rate of 0.0021 s^{-1}. The experimental design is shown in figure 1.

To study the continuous evolution in space and during the strain pattern, a speckle structure was created on the surface of the sample, which was obtained by applying the white and black finely dispersed contrastive aerosol paints to the surface of the sample. This allowed us to create a variety of contrasting irregularly located points on the surface.

The surface strain of a steel plate was measured by using a Vic-3D digital optical system based on the correlation method of digital stereoscopic images. By the process of testing, the VicSnap program allows to record synchronously the images from two cameras (figure 1), which are processed by the VIC-3D program. As a result, the geometrical parameters of the surface are calculated (*X, Y, Z* coordinates for each analyzed point), as well as the relative strains (ε_{xx} – along the *OX* axis, ε_{yy} – along the *OY*

3. Results and discussion

The application of the VIC-3D system allows obtaining the patterns, which elaborately reflect the continuous evolution in space and during the distribution of isofields of relative strains on the surface of steel plates, as well as the nature of the distribution of deformation fields at each stage of the stress-strain curve (figure 4-6).

The stage I of the stress-strain curve (figure 3) corresponds to the patterns of horizontal relative strains.

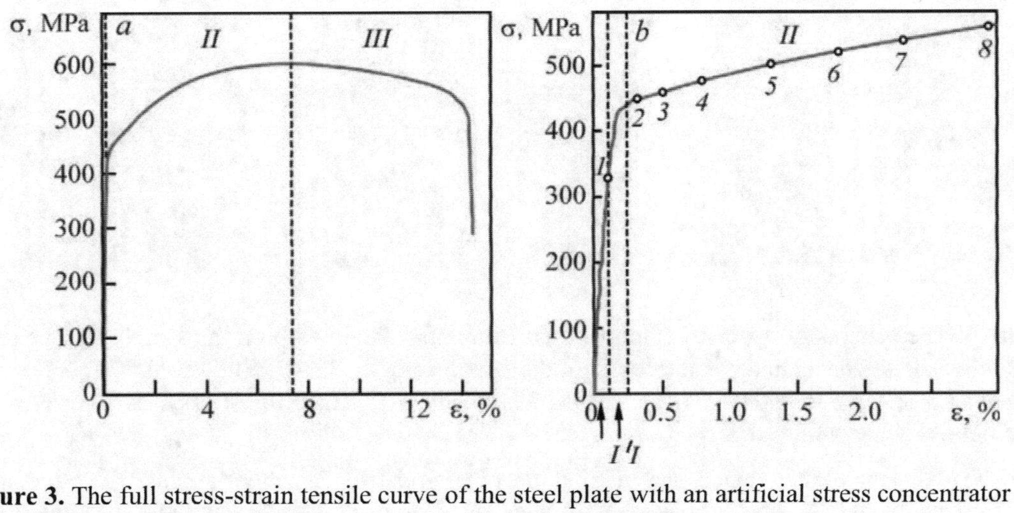

Figure 3. The full stress-strain tensile curve of the steel plate with an artificial stress concentrator (*a*) and a fragment of the stress-strain curve in the first strain stage (*b*).

ε_{XX}, in which the distribution of strain fields is characterized by the chaotic arrangement of the plastic compressive strain zones (figure 4, pattern (a)). The local zones of plastic strain are stretched along the axis of stretching (*OY* axis, figure 1). It corresponds to the strain geometry of the sample. During the stretching of the platelike sample, the elongation along the *OY* axis and the compression in the horizontal direction along the *OX* axis occur. Thus, the compression stresses along the *OX* axis lead to compression of the plastic strain zones. For this purpose, the zones of plastic strain are grouped into two wedge-shaped areas with the base of the wedges in the upper and lower edges of the plate and the tip of the wedges in the central part of the sample. In the neck formed by the undercuts between the stress concentrators, a deformation zone is formed by compression.

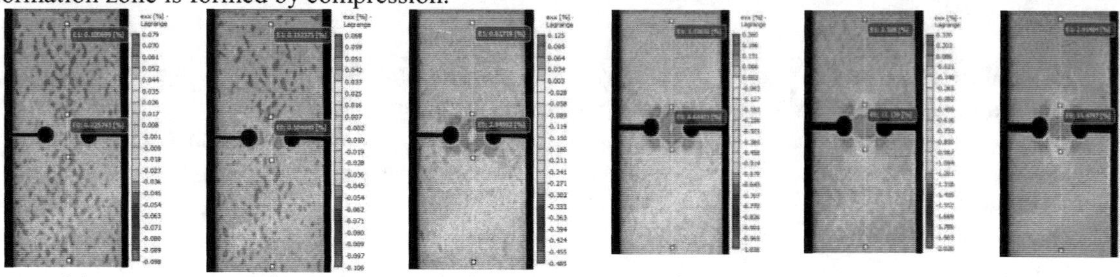

(a) (b) (c) (f) (g) (h)

Figure 4. The patterns of distributions of horizontal relative strains ε_{XX} on the sample surface of 09G2S steel with a stress concentrator in the center of the sample: (a)–ε = 0.10%; (b)–ε = 0.15%; (c)–ε = 0.61%; (d)–ε = 1.00%; (e)–ε = 2.33%; (f)–ε = 2.91%. The points (a)– (h) of the stress-strain curve in figure 3 correspond to these patterns.

In the patterns of the distributions of the relative longitudinal strains ε_{YY} along the tensile axis at the stage I, the zones of plastic strain are elongated along the *OX* axis, which is perpendicular to the tensile direction of the sample. In this case, in the local zone of the stress concentrator, the material is stretched with the local strain, which exceeds more than twice the magnitude of the highest average strain over the entire sample (figure 5, pattern (a)).

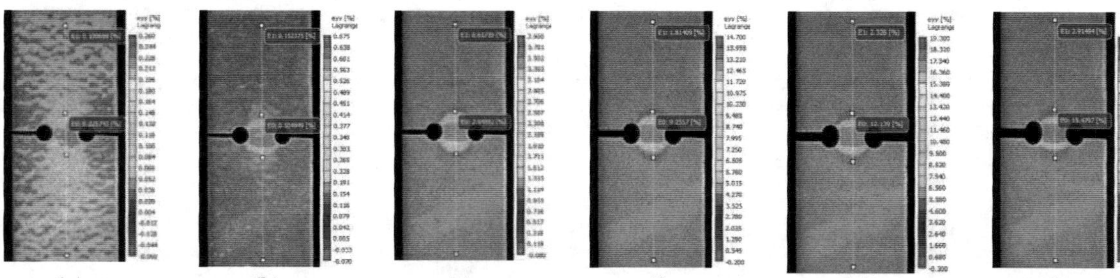

(a) (b) (c) (f) (g) (h)

Figure 5. The patterns of the distributions of vertical relative strains ε_{YY} on the sample surface of 09G2S steel with a stress concentrator in the center of the sample: (a)–ε = 0.10%; (b)–ε = 0.15%; (c)–ε = 0.61%; (d)–ε = 1.0%; (e)–ε = 2.33%; (f)–ε = 2.91%. The points (a)– (h) of the stress-strain curve in figure 3 correspond to these patterns.

When the sample is stretched, the distribution of the relative tangential strains ε_{XY} at the stage I is characterized by the chaotic arrangement of the zones of plastic strain, and the asymmetric one in reference to the sign of strain of the longitudinal and transverse axes of the sample (figure 6, pattern (b)).

At the stage I′ (figure 3) of elastic-plastic transition from the elastic to plastic strain in the patterns of horizontal relative strains ε_{XX}, a decrease of the density zones of plastic compression strain is observed (figure 5, pattern (b)). A combined zone of plastic strain, consisted of the tension field along the axis *OX*, is formed in the neck.

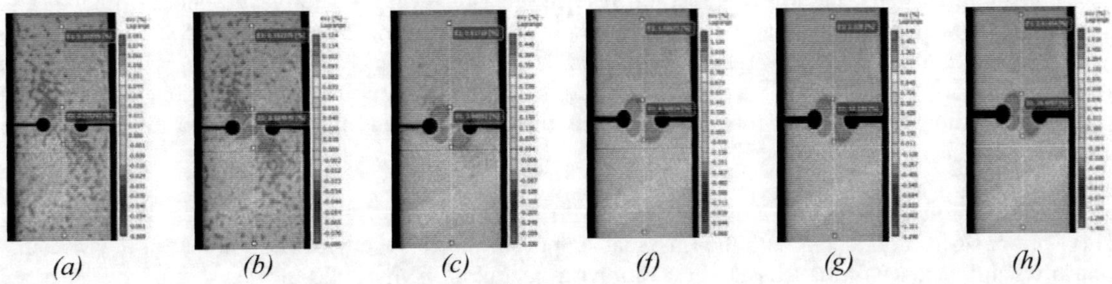

Figure 6. The patterns of the distributions of tangential strains ε_{XY} on the surface of the sample of 09G2S steel with a stress concentrator in the center of the sample: (a) $-\varepsilon$= 0.10%; (b)$-\varepsilon$ = 0.15%; (c)$-\varepsilon$= 0.61%; (d)$-\varepsilon$ = 1.0%; (e)$-\varepsilon$ = 2.33%; (f)$-\varepsilon$ = 2.91%. The points (a)– (h) of the stress-strain curve in figure 3 correspond to these patterns.

At the stage I′ (figure 3), in the patterns of the vertical relative strains εYY, one can observe the de-emphasis of the strain difference between the zones of plastic strain and background strain over the whole sample are with the exception of the zone of stress concentrator (figure 5, pattern (b)). In the neck formed by the undercuts, a tension field is observed, where the strain is more than 6 times higher than the average strain of the sample. At this stage, the system is practically in a weakly stable condition of the transition from the elastic strain to the plastic one. The importance of this stage is emphasized.

The distribution of the relative tangential ε_{XY} strains at stage I′ is characterized by a chaotic alternation of local zones of compressive and stretching plastic strains. In the neck at an angle of 45°, a zone of plastic strain is formed, made by the areas of compression and tension (figure 6, pattern (b)).

At the stage II of the stress-strain curve (figure 3), it can be seen that the whole plastic strain is localized in the zone of the stress concentrator and is characterized by the significant compressive strain (ε_{XX}<0) in the patterns of transverse relative strains ε_{XX}. In this case, the isofields are redistributed in the upper and lower parts of the sample, and it is characterized by the homogeneous strain with the value by an order less than in the zone of the stress concentrator. Almost complete disappearance of the plastic strain zones is observed (figure 4, patterns (c)-(f)).

The analysis of the strain patterns of vertical relative ε_{YY} strains at stage II shows that the whole plastic strain is localized in the zone of the stress concentrator and is characterized by the significant tensile strain (ε_{YY}> 0) with smaller and homogeneously distributed strains in the upper and lower parts of the sample (figure 5, patterns (c)-(f)).

A similar change in the behavior of the distributions of strain fields on the sample surface during the transition to stage II is observed in the distributions patterns of tangential strains ε_{XY} (figure 6, patterns (c)-(f)).

4. Conclusion

1. It is evaluated that each stage in the diagram of the stress-strain curve $\sigma = f(\varepsilon)$ of tension of the steel sample from 09G2S steel with a stress concentrator in the center of the plate corresponds to its own distribution of strain fields on the surface of steel samples. In this case, there is a continuous evolution in space and during the distributions of stress-strain states of structural elements in the zone of stress concentrators under the tension of the plate.

2. It was evaluated that the behavior of the change in the diagram σ ε is determined mainly by the steel strain in the local zone of the stress concentrator at the stage of plastic strain.
3. The elastic-plastic strain processes occur at the stage I′. The distinction of a narrow stage I′ on the stress-strain curve is due to the transition from the elastic strain to the plastic one. At this stage, the system is practically in a weakly stable condition of the transition from the elastic strain to the plastic one. The importance of this stage is emphasized.
4. The experimentally obtained strain diagrams and their corresponding representations and patterns of the distribution of isofields can be used to verify the finite-element models when performing numerical investigations.

5. Acknowledgments
The work has been conducted with the financial support of the Government Assignment of the Ministry of Education and Science of the Russian Federation (project No. FEMN-2020-0004).

References
[1] Demeshkin A G, Korenev V M and Kurguzov V D 2012 *Mech. Solids* pp 110-121
[2] Aim B and Nult S R 2010 *Experimental Mechanics* vol 50 pp 117-123
[3] Sutton M A, Orteu J J and Schreier H 2009 *Image Correlation for Shape, Motion and Deformation Measurements (*Columbia: University of South Carolina) p 364
[4] Tretyakova T V and Wildemann V E 2018 *Phys. Mesomechanics* **21** pp 314-19
[5] Tretyakova T and Zubova E 2018 *Procedia Structral Integrity* **13** pp 1739-44
[6] Ustinov A M, Klimenov V A, Klopotov A A, Abzaev Yu A and Ovcharenko V E 2018 *Materials Science Forum* vol 938 pp 62-69
[7] Ustinov A, Kopanitsa D, Abzaev Yu, Klopotov A, Koshko B, and Kopanitsa G 2017 *AIP Conference Proceedings* No 1800 p 040007 (1–6)
[8] Saraev Yu N, Golikov N I and Sidorov M M *Metal Working (Technology, Equipment, Tools)* pp 30–42
[9] Ustinov A M, Klopotov A A, Potekayev A I, Abzaev Yu A and Plevkov V S 2018 *Proceedings of the Altai State University* pp 58-69
[10] Kozlov E V, Koneva N.A. and et. al. 2004 *Phys. Mesomechanics* **7** pp 93-113
[11] Kozlov E V 2006 Structure and Resistance to Deformation of FCC Metals and Alloys *Severe Plastic Deformation: Toward Bulk Production of Nanostructured Materials* ed B S Altan (New York: Nova Publishers, Inc.) pp 295-332
[12] Panin V E., Elsukova T F and et. al. 2015 *Phys. Mesomechanics* **11**pp 1-7
[13] Panin V E, Panin A V and et. al. 2017 *Phys. Mesomechanics* **11** pp 57-71

Evaluation of strain-stress state of vertical tank reinforced by carbon tyre based on numerical researches in ANSYS PC

V A Vlasov[1], A A Klopotov[2], A S Plyaskin[3], V E Bunkov[4,7], A M Ustinov[5] and M E Savintceva[6]

[1]Rector of Tomsk State University of Architecture and Building, 634003, sq. Solyanaya 2, Tomsk, Russia
[2]Doctor of Physical and Mathematical Sciences, Professor, Tomsk State University of Architecture and Building, 634003, sq. Solyanaya 2, Tomsk, Russia
[3]Head of the Department "Metal and Wooden Structures", Tomsk State University of Architecture and Building, 634003, sq. Solyanaya 2, Tomsk, Russia
[4]Department "Metal and Wooden Structures", Tomsk State University of Architecture and Building, 634003, sq. Solyanaya 2, Tomsk, Russia
[5]Department "Metal and Wooden Structures", Tomsk State University of Architecture and Building, 634003, sq. Solyanaya 2, Tomsk, Russia
[6]Institute of International Relations and Internationalization of Education, Tomsk State University of Architecture and Building, 634003, sq. Solyanaya 2, Tomsk, Russia

E-mail: [1]rector@tsuab.ru, [2]klopotovaa@tsuab.ru, [3]Plyaskinandrei@mail.ru, [4]Viktorbunkov@inbox.ru, [5]ustinov_a_m@tsuab.ru and [6]msavintseva@yandex.ru

Abstract. The article presents the results of numerical studies of the stress-strain state of the models of a vertical cylindrical tank with corrosion wear and with the strengthening of the wall of the first ring by an external and internal composite tyre. Based on the results of the studies, the maximum percentage of corrosion wear of the wall of the first ring of the tank have been determined, and the dependences of the influence of the installation of the internal and external carbon composite tyre on the stress-strain state have been obtained.

1. Introduction

Nowadays, the greater part of the operating tank farm of steel tanks of the hydrocarbon process industry of Russia has exceeded the regulatory time limit. Long-term operation of capacitive equipment without timely overhaul results in defects that reduce operating reliability [1]. The current task is to develop new methods or adapt existing ones for the restoration of building structures, using modern materials with high operational characteristics.

The use of carbon composites in reinforcement of supporting structures is increasingly attracting researchers in this field every day [2, 3]. This phenomenon is primarily due to significant technological advances in the production of carbon fibers and synthetic binders. Adhesive bonding of metals to other materials becomes a very reliable method of joining elements in items and structures, and has a number of advantages over other types of bonding. It should be noted that riveted, bolted and welded joints have an uneven distribution of stresses at the junction, and are also weakened by holes under rivets and bolts, and increase the weight of structures. The bonding of metals and alloys by welding results in internal stresses in the weld zone and decrease in heat treatment strength. Adhesive joints have uniform stress distribution in bonded materials and are made by unfired method.

The works [4-7] show the results of tests of thin steel plates, reinforced with carbon plastic layers at cyclic quasi-static load. The authors conclude that the use of carbon plastic laminates can significantly increase yield strength, ultimate strength and stability. Fiber orientation is an important factor in shear enhancement. The greatest shear strength and resistance can be obtained by applying the basic orientation of carbon plastic laminates along the direction of the stress fields. When the thickness of the adhesive changes, the failure mode changes from cohesive failure to interfacial destruction of the adhesive on the steel surface. For cohesive fracture bonded joints, the maximum load increased as the adhesive thickness increased from 1 to 2 mm.

2. Analysis of defects, arising during operation of tanks.

Statistics show that the main cause of the failure of oil tanks is the corrosive wear of the surface, coming into contact with the corrosive (figure 1) [8.9].

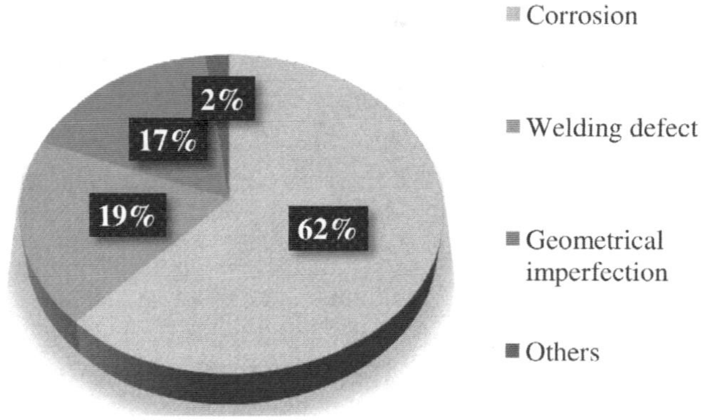

Figure 1. Statistics of VST failures

The analysis of safety expert review showed that the most of the corrosion defects are located on the bottom, corner weld joint and the first ring of the tank wall, which is confirmed by the frequent replacement of elements during overhaul repairs. The highest percentage of corrosion wear in the wall thickness was observed at the height of up to 30 cm from the level of the corner weld joint (Figure 2). The main cause of corrosion damage is the presence of bottom water at these levels. According to regulatory documents, GOST R 51858-2002 [10] in particular, the mass fraction of water is not more than 0.5% for the first and the second groups of oil, and not more than 1.0% for the third group.

The numerical research program provides three design models of tanks:

– vertical stock tank (VST) with corrosion wear of the inner surface of the first wall ring on a significant length in the area of abutment to the bottom. The nature of corrosion is the groups of shells that turn into continuous strips, as well as point depressions of the pitting type.

– VST with one inner tyre with the height of b = 1300 mm, made of carbon unidirectional tape Tape 230 on epoxy binder Resin 230.

– VST with two tyres, the inner one with the height of b = 1300 mm and the outer tyre with the height of b = 300, located at the height of 500 mm from the corner weld joint and made of carbon unidirectional tape Tape 230 on epoxy binder Resin 230.

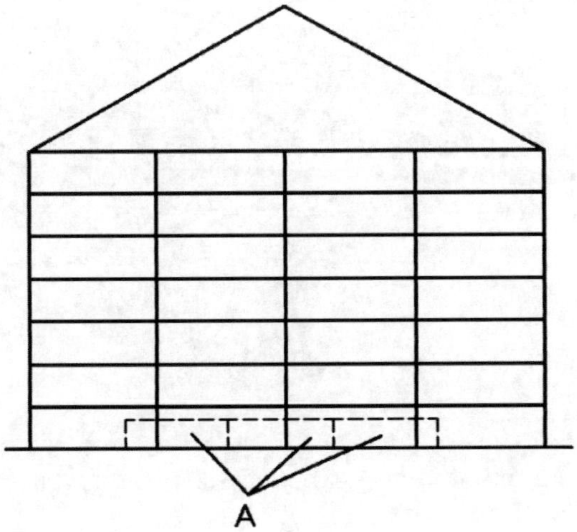

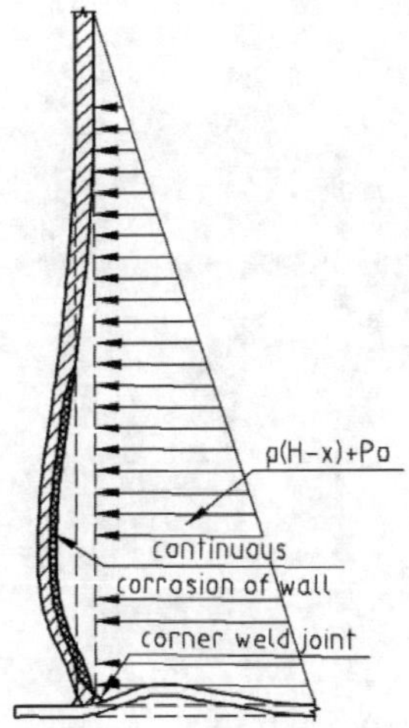

Location of defect Diagram of wall and bottom deformation

Figure 2. Arrangement of continuous corrosion on the inner surface of the tank wall.

The efficiency of the device of one external steel shroud ring in restoring the bearing capacity of the tank is justified in the work of M.A. Tarasenko [11].

According to [12-14], the maximum effective stresses of σ_{max}, MPa, in the tank wall should not exceed the values, specified in Table 1.

Table 1. Maximum allowable stresses in tank elements.

	For lower wall ring, MPa	For other rings of wall with bottom, MPa	For junction, MPa
Maximum allowable stresses under operating conditions	176,67	201,90	302,86

3. Basic data for creating tank simulation models.

To perform VAT calculations, the ANSYS software-computing complex is used, which allows:

- creating FE model and determine stress-strain state of tank structures: tank wall, bottom, welded joints;
- setting distributed, local, hydrostatic and inertial loads;
- setting loads from previously applied stresses (welding stresses);
- solving elastic (linear) and elastic-plastic problems (nonlinear);

To build a geometric model of the tank, a typical design of RVS 5000 m3 “TP 704-1-27” was adopted [15].

The finite element model of the tank was constructed, using shell elements SHELL181, having a number of features, inherent in thin-walled shells. Face meshing is used to create a uniform ordered mesh on the surface of the tank wall. The size of the finite element mesh was 0.05 m (square side). In order to improve the accuracy of calculations in the wall and bottom interface, the FE grid was thickened 4 times, using the Refination function [16].

Statistical information of finite element tank model is given in Table 2.

Table 2. Statistical information of the finite element model.

	Total
nodes	332361
elements	326905
material properties	2

IOP Conf. Series: Materials Science and Engineering **911** (2020) 012009 doi:10.1088/1757-899X/911/1/012009

Hydrostatic pressure (figure 3) and overpressure (figure 4) were used as tank loads.

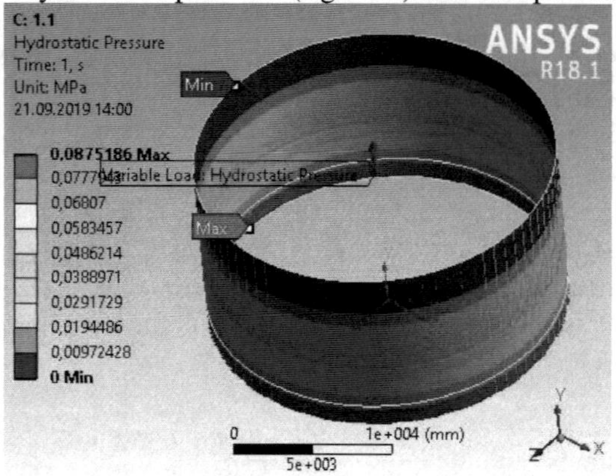

Figure 3. Hydrostatic pressure on VST wall from stored product.

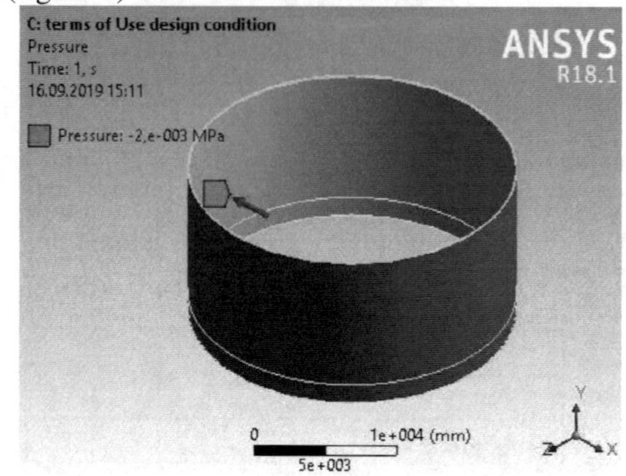

Figure 4. Excessive internal pressure on VST wall.

4. Results of numerical calculation.

Based on the results of the conducted studies, the dependence of the strains αeq on the value of corrosion wear of the wall of the first ring in the range from 0 to 40% was obtained with reinforcement by carbon composite tyres and without reinforcement (figure 5). There is also a diagram of the dependence of deformations (wall deviation from vertical) of the tank wall on the value of corrosion wear for three design models.

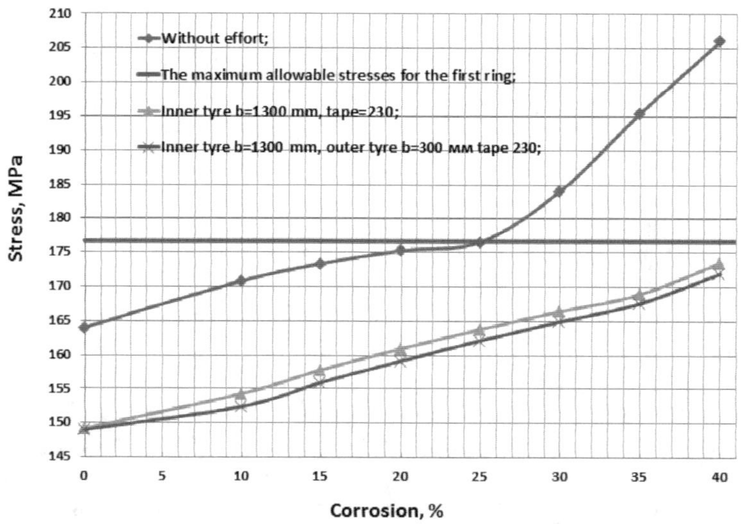

Figure 5. Stresses in the first wall ring at corrosion value from 0 to 40%.

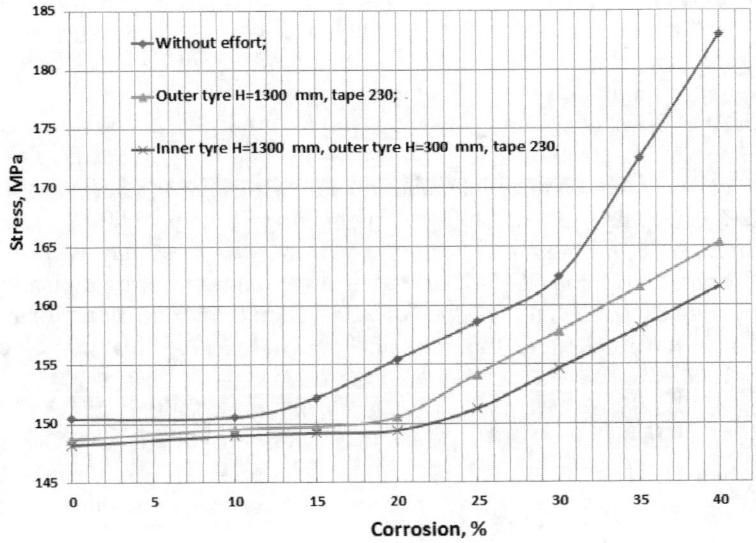

Figure 6. Deformations in the tank wall at corrosion value from 0 to 40%.

Analysis of the obtained data showed that the excess of permissible stresses occurs with continuous corrosion of the first wall ring, exceeding 25%.

Installation of one internal carbon composite shroud, with corrosion of 25%, allows to reduce the level of maximum stresses in the wall by 7.25% and deformations by 1.25%

Combined installation of internal and external carbon composite shrouds with similar corrosion percentage showed a decrease in wall stresses by 8.27% and deformations by 1.625%.

5. Conclusions.

The application of the internal carbon composite shroud can be effectively used in restoring service suitability and extending the life cycle of tank with corrosion wear of not more than 40%.

The use of a combined device of internal and external carbon composite shrouds does not significantly reduce the equivalent stresses, arising in the walls of the first ring.

References

[1] Gaisin E SH and Gaisin M SH 2016 Current state of problems of ensuring the reliability of oil and oil products tanks *Transportation and Storage of Petroleum Products and Carbon Raw Materials* **2** pp 31-40.

[2] Lu Y, Li W, Li S and X. Liand X and Zhu T 2015 Study of the tensile properties of CFRP strengthened steel plates *Polymers* **7** pp 2595-2610.

[3] Majidi H R, Razaviand S M J and Berto F 2017 Failure assessment of steel/CFRP double trap joints *Metals* **7** p 255.

[4] Poul M K, F Nateghi-Alahi and Zhao X L 2016 Experimental testing on CFRP strengthened thin steel plates under shear loading *Thin–Walled Structures* **109** pp 217-226.

[5] Wang H, Wu G, Dai Y and He X 2016 Determination of the bond–slip behavior of CFRP to steel bonded interfaces using digital image correlation *Journal of Reinforced Plastics and Composites* **35(18)** pp 1353-67.

[6] Wang B, Hu X and Lu P 2017 Improvement of adhesive bonding of grit-blasted steel substrates by using diluted resin as a primer *International Journal of Adhesion & Adhesives* **73** p 92-99.

[7] Dawood M and Rizkalla S 2010 Environmental durability of a CFRP system for strengthening steel structures *Construction and Building Materials* **24** pp 1682-89.

[8] Abdullin I G, Gareev A G and Mostovoy A V 1997 *Corrosion-mechanical Resistance of Oil*

and Gas Pipeline Systems. Diagnosis and prediction of durability (Ufa: ed Tilem) p 111.

[9] Vengertsev V A and et al 1986 Analysis of Failures of Metal Tanks at Oil Products Facilities. NTIS *Scientific and Production Achievements of the Petroleum Industry in the New Economic conditions* **6** pp 3-4.

[10] *GOST R 51858-2002 2006 Oil. General Specifications* Input 30.06.2002 (Moscow: Standard Publishing House) p 28.

[11] Tarasenko M A 2012 *Development of methodology for restoring the load capacity of tank with corrosion damage* Thesis Paper of Ph.D. in Engineering Science 25.00.19 (Tyumen) p 155.

[12] *GOST 31385-2016 2016 Vertical cylindrical steel tanks for oil and oil products General Technical Specifications* Input 2017-03-01 (Moscow: Standartinform publishing house) p 97.

[13] VSP 34-01-03 / Ministry of Defence of the Russian Federation Guidelines for the Calculation and Design of Metal Tanks and Pipelines in Fuel Storages of Ministry of Defence Input 07.04.2003 p 160.

[14] *SP 16.13330.2016 2011 Steel Structures.* Updated Version of SNiP II-23-81 *" (as amended, with Change N 1) Input 28.08.2017 (Moscow: Ministry of Regional Development of the Russian Federation) p 177.

[15] *Typical project 704-1-27 Steel Tanks for Petroleum Products for Low Temperature Operation*. 5000 cu m Tank.

[16] Tarasenko A A and et al 2013 Tank model in ansys workbench 14.5 *Basic Research* **10** pp 3404-08.

Role of temperature factor in the work of a cellular reinforced concrete ice-resistant platform

V A Zimnyukov[1] and M I Zborovskaya[1]

[1]Russian State Agrarian University - Moscow Timiryazev Agricultural Academy, Moscow, Russian Federation

E-mail: zimnyukov@rgau-msha.ru

Abstract. The article presents materials reflecting the results of calculations by the finite element method on a computer (according to a special program) of the thermal state of the cellular fragment of the reinforced concrete support block LSP taking into account the influence of seasonal and decadal outdoor temperatures, as well as the effect of changes in the thermal characteristics of the backfill soil on the thermal state of the ice-resistant stationary platform . This made it possible for the first time to evaluate the joint work of a thin reinforced concrete frame and a massive body of backfill soil under temperature effects. In the case of the operation of the cellular structure without the backfill soil, seasonal and ten-day temperature fluctuations create a more unfavourable temperature regime in comparison with the cellular structure with the backfill soil. In the absence of backfill soil, negative temperatures penetrate most deeply into the thickness of the structure. The presented data can also serve as the basis for a preliminary assessment of the thermal state of LSP in the form of a fully cellular reinforced concrete structure and in the future to determine its thermally stressed state. Only considering the combined effect of ice, wave and temperature loads can we really evaluate the operation of such a complex structure in terms of its reliability and safety.

1. Introduction

It is known that the most numerous groups of structures used on the sea shelf are used for oil and gas field structures. Depending on their mobility, they are divided into stationary and mobile. Stationary platforms are of gravitational type and often pile construction.

Most platforms are erected from metal structures, whose service life in sea water is 2–3 times shorter than on land [1-4].

Another variety of gravity platform are reinforced concrete gravity platforms with the partial use of cells filled with sand to increase the stability of the structure.

Unfortunately, to date, little attention has been paid to the work of structures on temperature effects. It is proposed to take into account the behaviour of the structure in the zone of variable water level [5-7], however, having zero degrees in the underwater part and up to -50 ° C in the surface part, the whole spectrum of the annual course of outdoor temperatures (including seasonal and ten-day temperature fluctuations) should be taken into account. At the same time, it is necessary to consider the possibility of contact between the produced hot oil and the walls of the structure and the significant temperature difference created in this structure. Regulatory documents of the 1980s (BCH 41.88) classify temperature effects as special loads along with seismic, wave and ice loads.

We propose to consider a fully cellular design of the platform with a concrete cellular frame filled with backfill soil (Figure 1). It is economical, more resilient, fire resistant and most effectively absorbs vibration created by equipment. On the other hand, the use of cellular concrete (or reinforced concrete) structures with backfill soil, as well as hydroelectric power station floating blocks, is also justified by the following criteria: safety of construction and installation works, low operating costs, high construction speed (including fast transportation), high reliability and the strength of structures under the influence of all external forces, including special (in emergency situations) and various temperature effects. The cellular structures of such structures can be built in various climatic and geological conditions (on rocky and non-rocky foundations), and any local soil of useful excavations can be used as backfill soil for cells. According to [8-10] "reinforced concrete is the most acceptable material of support blocks for work in seismic areas and in emergency situations".

Unfortunately, a few issues related to the design and calculations of these structures did not find a sufficiently complete justification. Currently, there are several approximate calculation methods based on various assumptions.

There are practically no calculations that implement the joint work of the concrete frame of the cellular structure and the backfill soil, taking into account the temperature effects.

At ONIL "Strength and Sustainability" of the Department of Hydrotechnical Structures of the Institute of Land Reclamation, Water Management and Construction named after A.N. Kostyakov Russian State Agrarian University - Moscow Timiryazev Agricultural Academy for a number of years carried out model studies and computer calculations of cellular structures of hydraulic structures taking into account the influence of backfill soil and base.

As the results of earlier laboratory studies have shown, various temperature effects have a significant impact on the operation of such structures. The objective of this work was to determine the thermal state of the cellular structure from seasonal and ten-day fluctuations in outdoor temperatures, taking into account changes in the thermophysical properties of soils.

2. Materials and methods

The cellular type structures [9] are a reinforced concrete frame consisting of mutually intersecting radial and annular bulkheads made up of flat fragments that form cells filled or unfilled with soil (Figure 1).

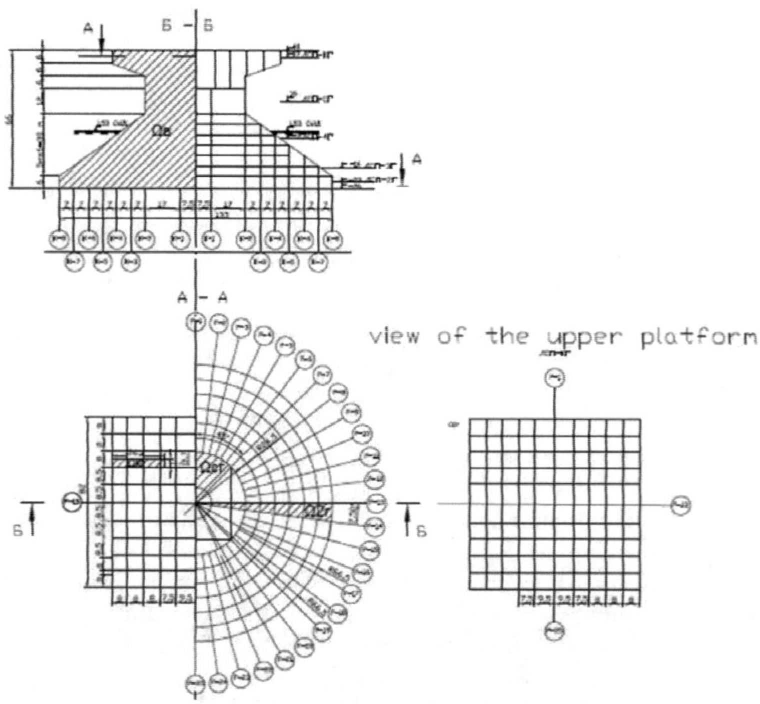

Figure 1. The main parameters and design sections of the ice-resistant stationary platform (LSP). Support block, axial alignment lines and planes (radial bulkheads with even numbers and ring bulkheads with odd numbers are made from panels with holes).

The cellular structures of hydraulic structures are classified as gravitational shear. Existing computer software packages that implement the finite element method, as applied to the calculations of cellular structures, do not fully reflect the physics of the phenomena occurring in such structures. Great difficulties arise when considering the joint work of a thin reinforced concrete frame and a massive backfill body. The physics of the phenomena and its mathematical description regarding the joint work of the base soil, the backfill soil, and the carcass are not completely clear.

The numerical studies of the thermal state of the cellular structure presented in this paper were carried out according to the Quasis program. The objective of this work was to determine the thermal state of the cellular structure from seasonal and ten-day fluctuations in outdoor temperatures, considering changes in the thermophysical properties of the backfill soil.

When determining the thermal regime of the cellular structures of the LSP, we considered the effect of ambient temperatures (water, air) from the front wall for a series of flat horizontal sections and from the front and top walls for the vertical section of the LSP (LSP-V) (Figure 1).

The temperature regime of the honeycomb structure under consideration is formed under the influence of heat exchange with the environment at the borders of the selected computational domain (Figure 1). To determine the thermal regime of a cellular structure, it is necessary to have physical and mechanical and thermophysical characteristics of the carcass material (concrete or reinforced concrete) and the soil of the cell backfill.

The thermophysical characteristics of the frame and the soil of the cell backfill determine the intensity of the change in the temperature field of the structure in time and in coordinates.
Physical and mechanical and thermophysical characteristics of the materials of the construction considered in the calculations are shown in Table 1.

Table 1. The main thermophysical and physical and mechanical characteristics of the materials of the frame and filling the cells LSP[a]

Material	Thermophysical and physical and mechanical characteristics of materials of the computational domain							Notes
	$\lambda_x = \lambda_y$ Вт/м*К	$\alpha_{п}$, Вт/м^2*К	c, кДж/кг*К	E, МПа	μ	γ, Н/м^3	α, 1/ degree	
Concrete	2,55	Concrete-air 23,26 Concrete - water 232,6	1,0048	29000	0,15	24000	$1*10^{-5}$	
Backfill cells (sand)	0,58	-	1,0006	26	0,235	18000	$1*10^{-5}$	
No cell filling (air)	0,58	-	1,0006	-	-	18000	$1*10^{-5}$	Simulated air gap e→0

[a] λ_x и λ_y, c, γ – vary with moisture and material density

To carry out the calculations, the real annual course of average monthly temperatures was brought to a harmonic form $t = t_{ср.г.} + A_г \cos(\omega\tau + \varepsilon_г)$ degrees Celsius, where

t - the temperature of the external environment at time τ, degrees Celsius;

$A_г$ – the full amplitude of the fluctuations in the monthly average ambient temperatures relative to $t_{ср.г.}$, degrees Celsius;

$t_{ср.г.}$ – average annual temperature, degrees Celsius;

τ – time, counted from the beginning of the year, hours;

$\tau_{год.}$ – period, number of hours in a year, τгод/ = 8760 hours;

$\varepsilon_г$ – is the initial phase of harmonic oscillations;

$\varepsilon_г$ = hours = 0.2619 radians;

IOP Conf. Series: Materials Science and Engineering **911** (2020) 012010 doi:10.1088/1757-899X/911/1/012010

$\omega_г = {}^{2\pi}/_{\tau_{год.}}$ – cyclic frequency,

$\omega_г$ = 0,00071726 radians / hour.

Thus, the calculated harmonic temperature fluctuations of the external environment have the following form with the seasonal course of temperatures:

а) outdoor air

t=-3-16.25 cos (0.00071726τ-0.2619) degrees Celsius;

б) water on the horizon 0.0m

t=2.88-6.35 cos (0.00071726τ-0.2619) degrees Celsius;

в) water on the horizon 10.0m

t=1.7-4.875 cos (0.00071726τ-0.2619) degrees Celsius;

г) water on the horizon 20.0m

t=0,91-3.45 cos (0.00071726τ-0.2619) degrees Celsius/

3. Results

As a result of the thermal calculation according to the calculation schemes described above, the thermal fields were determined for 12 months on the 15th day of each month for the annual temperature course and for 12 steps of 20 hours for a ten-day temperature course in a quasi-stationary setting of the ambient temperature course.

Figure 2 shows the effect of the annual temperature variation on the distribution of temperature fields in a structure in January over horizontal sections, and in Figure 3 - in April over horizontal sections. As can be seen from Figure 2, seasonal fluctuations in outdoor temperatures in January have a significant effect on the formation of the thermal state of the upper part of the platform (from the 28.00 m mark to the 40.00 m mark). their most significant effect is felt in its cantilever part, and complete attenuation occurs only inside the middle part of the structure (in the "glass") at the end of the fifth from the front wall of the cell (along the P-I axis).

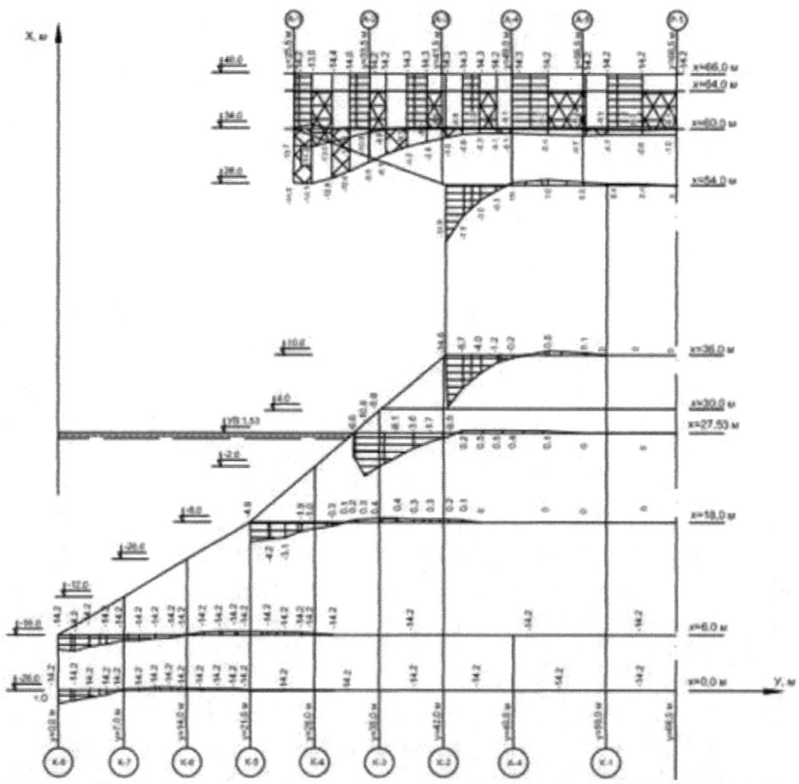

Figure 2. The effect of the annual temperature variation on the distribution of temperature fields in a structure in January over horizontal sections (the water level is located at 1.53 m).

In this case, at around 28.00 m at the end of the third from the front wall of the cell (along the axes A3 and A4), the transition of negative temperatures to positive is observed.

As can be seen from Figure 3, in April, a certain effect of seasonal fluctuations in the temperature of the outside air (taking into account fluctuations in water temperature) on the thermal state of the structure as a whole and also the most significant on its upper part is again noted. The above results show that the cantilever part of the LSP is most susceptible to adverse effects of temperature.

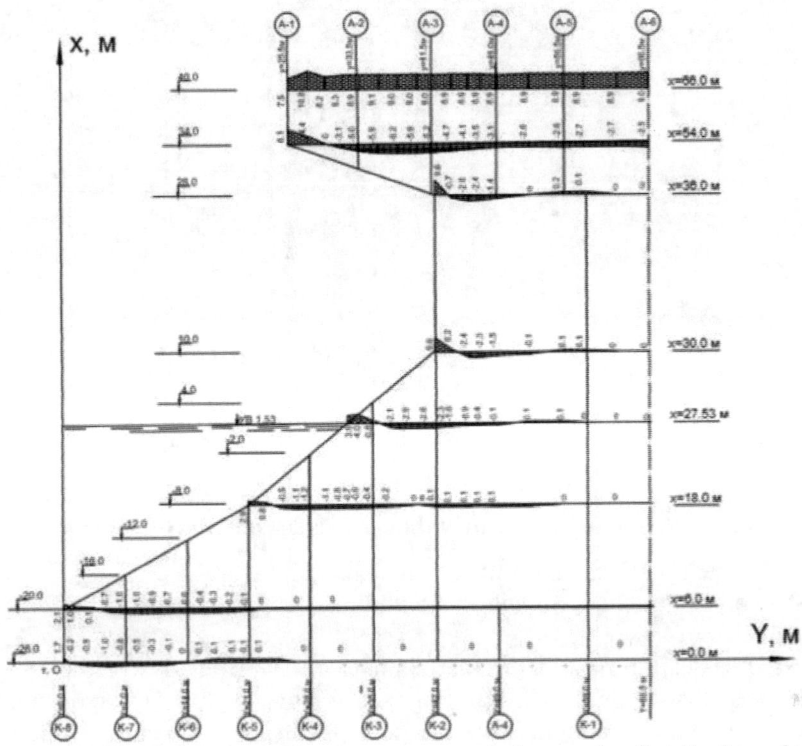

Figure 3. The effect of the annual temperature variation on the distribution of temperature fields in a structure in April over horizontal sections.

The temperature distribution obtained by horizontal sections allows you to fully evaluate the thermal state of the specified design of the cellular type.

From the temperature distribution shown in Figure 3 in April, the temperature regime established in the structure is somewhat more favourable than in January. An exception is the cantilever part at 34.0 m and 28.0 m, where, because of a delay in temperature over time, negative temperatures penetrate deeper into the thickness of the structure. At around 34.0 m, they are marked along the entire perimeter of the structure.

At levels of 1.53 m and -8.0 m in April, negative temperatures also penetrate deeper into the structure (about one cell). The same pattern appears at the base.

The above results show that the cantilever part of the LSP is most susceptible to adverse effects of temperature. It is a prismatic structure of complex shape (see figure 1).

The influence of decadal temperatures on the distribution of temperature fields in a structure/

The temperature distribution in the building, respectively, in the fourth part of the decade (4th 20 hours) for a few vertical sections is presented in Figure 4.

From the temperature diagrams shown in Figure 4, it can also be noted that the effect of decade temperatures is limited by the cantilever part of the LSP (within two cells along the axes A - 1, A - 2, and A - 3).

From the mark of 28.0 m to the mark of 10.0 m, the decadal temperatures decay at less than 1/3 of the size of the cells located at the front surface.

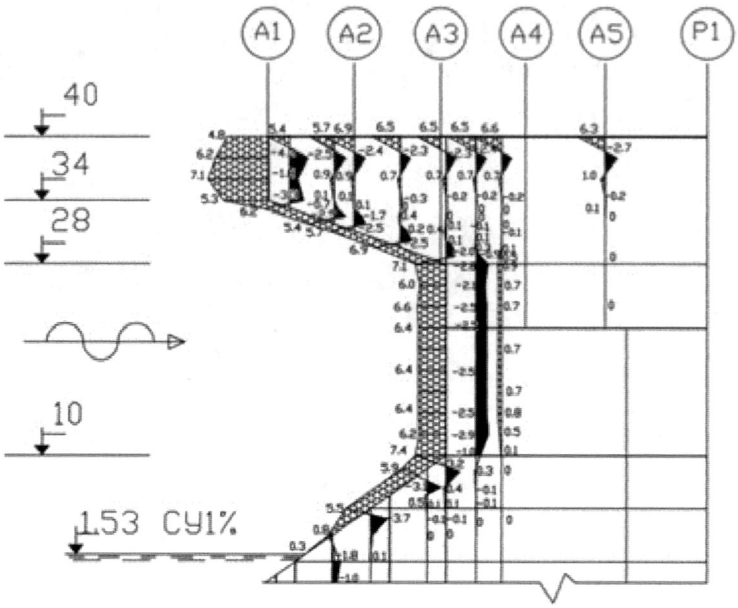

Figure 4. Influence of the ten-day course of temperatures on the distribution of temperature fields in a structure in the fourth 20 hours of a decade over vertical section.

The presence of water (mark 1.53 m) almost completely removes the effect of ten-day temperatures on the thermal state of the structure. So below the -2.0 m mark, a constant temperature field is almost fixed.

The paper also assesses the role of the thermophysical characteristics of the backfill soil in the formation of the thermal state of the cellular fragment. Figure 5 shows the temperature distribution over the LSP cross section at around 16.0 m (figure 1) from the seasonal outdoor temperature fluctuation in the backfill soil, depending on the soil thermal conductivity coefficient λr for the amplitude A_r = 1 degree in January.

As can be seen from Figure 5, in January, with a decrease in the value of λr, a decrease in the zone of negative temperatures in the soil is noted, which shifts to the front surface (curve 2). In this case, curve 1 corresponds to the temperature distribution along the concrete wall of the cellular structure of the LSP.

The absence of backfill soil (curve 3) negative temperatures penetrate deep into the thickness of the structure.

The curve numbers correspond to the variant numbers in table 2, column 1.

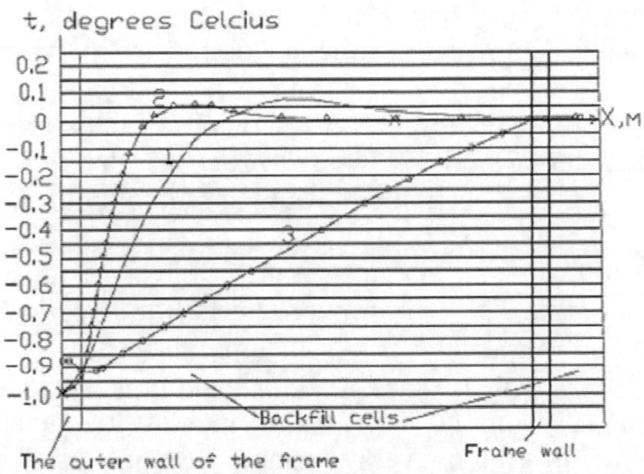

Figure 5. The influence of the thermal conductivity coefficient λ soil filling cells LSP on the temperature distribution in January over the section "wall - soil" (mark 16.0 m in figure 1) with seasonal fluctuations in outdoor temperature t = -1.0 cos ωτ, degrees for $t_{cp.r.}$ = 0.0 degrees.

Table 2. The main thermophysical and physical and mechanical characteristics of the materials of the frame and soil of the backfill of the cellular structure of the LSP according to the options [b]

Variant of the curve in the graph of Figure 5	Material	Thermophysical and physical and mechanical characteristics of materials of the computational domain				
		$\lambda_X = \lambda_Y$ Вт/м*К	$\alpha_п$, Вт/м^2*К	C, КДж/кг*К	γ, Н/м^3	α, 1/ degree
1	2	3	4	5	6	7
1	Concrete	2.56	23.26	1.0048	24000	1*10^{-5}
2	Concrete	2.56	23.26	1.0048	24000	1*10^{-5}
	Backfill (sand)	0.58		1.0006	18000	
3	Concrete	2.56	23.26	1.0048	24000	1*10^{-5}
	There is no backfill	0.58		0.0008	18000	

[b] λ_X and λ_Y , c, γ –vary with moisture and material density

4. Findings

1. New physical ideas about the behaviour of the structure of the type under consideration under temperature effects are obtained. As the real calculations on the definition of the thermal state of the soil have shown of a fragment working together with the backfill soil, along with seasonal fluctuations in the temperature of the outside air, it is necessary to take into account the ten-day fluctuations in the temperature of the outside air. The presence of large temperature differences between the front wall and adjacent bulkheads can lead to the appearance of significant values of temperature stresses.

2. The use of soil cells with different values of the coefficient of thermal conductivity as a backfill primarily affects the thermal regime of the soil massif (changes the values of temperature amplitudes and the degree of their attenuation along the depth of the structure). To a lesser extent, this affects the thermal state of the reinforced concrete frame.

3. In the case of the filling of the cellular fragment without soil, seasonal and ten-day temperature fluctuations create a more unfavourable temperature regime inside the structure compared to the structure filled with soil. In the absence of backfill soil, negative temperatures penetrate most deeply into the thickness of the structure.

4. The presented data can serve as a sufficient basis not only for a preliminary assessment of the thermal state of the specified structure, but in the future to determine its thermally stressed state.

5. In sections of the structure located below the water edge, insignificant temperatures are observed and their attenuation along the depth of the structure occurs much faster than in sections that are directly affected by outside air.

6. The combination of the properties of the floating blocks of hydroelectric power stations built in the regions of Western and Eastern Siberia allows us to conclude that the efficiency of the delivery of these cellular structures to any given location ensures high reliability of the structure according to the main parameters.

7. According to the current regulatory documents, the following gradation of temperature loads for LSPs can be given: 1) temporary loads - temperature effects (for a year with an average amplitude of average monthly temperatures); 2) special loads — external influences with maximum parameters — temperature effects determined for a year with a maximum amplitude of average monthly temperatures or a maximum low amplitude of average monthly temperatures; decade-long outdoor temperature fluctuations with a maximum average temperature per decade. These temperature effects are mandatory for design considerations.

References

[1] Musabirov A A 2013 *Razrabotka i issledovaniye primenimosti novoy konstruktsii ledostoykikh platform na melkovodnom arkticheskom shel'fe* Abstract for the degree of Ph.D (Moscow: National University of Oil and Gas "Gubkin University") p 24

[2] Vyakhirev R I, Nikitin B A and Mirzoev D A 1999 *Obustroystvo i osvoyeniye morskikh neftegazovykh mestorozhdeniy* (Moscow: Academy of Mining Sciences Press) p 373

[3] *SP581.3330.2012. Hydraulic structures. Basic statements*. Updated edition of SNiP 33-01-2003 (Moscow) p 39

[4] SP38.13330.2012. *Loads and impacts on Hydraulic structures (from wave, ice and ships)* SNiP2.06.04-82* (Moscow) p 91

[5] Finagenov O M 2013 *Scientific-Technical collection book Vesti Gazovoi Nauki* **3(14)** pp 162-166

[6] Kozachinsky Yu S and Gavrilenko V A 2017 *Hydrotechnika* **3 (48)** pp 30-33.

[7] Ilyina S V and Gnidov K P 2014 *Materials of conference. Modern high technologies* **5** pp 138-139

[8] Zimnyukov V A and Zborovskaya M I 2018 Nagruzki i vozdeystviya na ledostoykiye statsionarnyye platform *Preprint*

[9] Zimnyukov V A and Zborovskaya M I 2018 *Proceedings of the Academy of Problems of Water Management Sciences. Issue 12. Actual problems of water management construction* (Moscow: Lomonosow Moscow State University) pp 158-159

[10] Zimnyukov V A, Zborovskaya M I and Zaitsev A I 2017 *Uchot temperaturnykh vozdeystviy pri rabote gidrotekhnicheskikh sooruzheniy: uchebnoye posobiye*(Moscow: Russian State Agrarian University – Moscow Timiryazev Agricultural Academy) p 112

IOP Conf. Series: Materials Science and Engineering **911** (2020) 012011 doi:10.1088/1757-899X/911/1/012011

Algorithm for designing building constructions expressed by nonlinear functions

S N Volkova[1], A V Shleenko[2] and E E Sivak[1]

[1]Kursk State Agricultural Academy, Kursk, Russian Federation, [2]Southwest State University, Kursk, Russian Federation

E-mail: shleenko77@mail.ru

Abstract. The aim of the work is to develop an algorithm for designing building constructions, namely their geometric construction (the geometrical location of points) in the form of an expression using the formula (equation) of geometric property of construction expressed in the projection by nonlinear functions. The article gives an algorithm for finding analytical dependencies of nonlinear functions used in applications. You must select a coordinate system before composing the curve equation. The type of the desired equation depends on the choice of the coordinate system. There are no rules that could guide one in choosing a coordinate system, and the ability to make the most rational choice is given only by experience. In this paper, the examples of the choice of coordinate systems show the form of functions, namely, their simplification. In some cases, geometric construction is the only way to present a graphic image of the building structure itself, no matter how complex it may be in the Cartesian coordinate system. The polar coordinate system is presented here as an alternative for unique structures and structures of increased complexity in the Cartesian coordinate system. This is achieved by establishing a relationship between the current coordinates of the point located on the projection of the construction and constants characterizing the properties of this construction, called a function on the plane and a surface in space in researches. The prospect of this approach has proven itself: in the construction of exquisite palace temples, beautiful architectural ensembles that make the space surrounding a person comfortable.

1. Introduction.

Developing the ideas of V.G. Shukhov, the modern world gets innovative solutions where it is necessary to use not monoliths, but connecting links, like a radio mast in Moscow, Shukhov towers in the form of oil and gas pipelines, water tanks, domed roofs, etc. Therefore, the development of design algorithms for strong and lightweight building constructions due to reviving methods is especially relevant in the era of the digital economy, when increasing the accuracy of calculations of complex structures expressed by nonlinear functions allows you to save unaccounted materials and make future architecture projects strong and beautiful, with predetermined properties.

It is known that the complexity of the task of the designed construction depends on the choice of the coordinate system. Even in the case of choosing a polar coordinate system, the type of structure and its projection will depend on the choice of the location of the pole and the polar axis in it. For example, if the construction is represented by a circle in the projection, then the equation of the circle has a simpler form if the pole coincides with the center of the circle than if it is on the circle itself, namely $\rho = a$ and $\rho = 2a \cdot \cos \varphi$, respectively [1].

But in the case of a projection of a construction in the form of an ellipse, on the contrary, the view will be simpler for second-order curves if the pole is in one of the focal points of the ellipse and the polar axis passes through the foci. Thus, in polar coordinates, the same line can be determined by both an algebraic and a transcendental equation, depending on the choice of the pole and the polar axis. As a result, we cannot classify construction lines on the basis of their equations in polar coordinates, but these lines can be used to obtain a more convenient form compared to the Cartesian coordinate system in which lines such as the Archimedes spiral and the logarithmic spiral, as well as the four-petal rose, are specified by implicit functions. There is no algorithm for choosing a coordinate system, only experience allows us to make the most convenient choice.

2. Methods.

The widespread use of building constructions made up of metal beams arranged along rectilinear generators, combining high strength and lightness has determined their wide distribution in our country and abroad. The research method consists in formation of an equation of the desired construction, we choose an arbitrary point M with x and y coordinates lying on the desired curve in projection onto the Oxy plane. This means that x and y will be the current coordinates. To make a geometrical place of the points lying on the curve and satisfying a certain condition means to find the equation of this curve i.e. to express the geometric property of the curve using the formula. We have found the following equations using this method: Cassini oval, for the case m> n, as well as Bernoulli lemniscates, for the case m = n in Cartesian and polar coordinate systems on the plane, as well as their cylindrical surfaces in space with the possibility of their hardening along rectilinear generators Other paragraphs are indented (BodytextIndented style).

3. Results and Discussion.

We begin the presentation of the material with the statement of tasks. According to the developed algorithm, you need to find the equations of geometric location of points of the building construction by the condition: the product of distances for which up to two given points A and B is a constant value equal to m2. The distance between A and B is 2n. It is necessary to consider the construction option for the case when m = n and find building constructions that meet the specified conditions and build them.

To solve the above problems, namely, compiling the desired projections onto the Oxy plane, it is convenient to take the straight line passing through the A and B points through the A axis (then the coordinates of the A and B points will be simple), and take the straight line bisecting segment AB as the oy axis (equal points of points A and B allow us to assume the symmetry of the curve) (Figure 1).

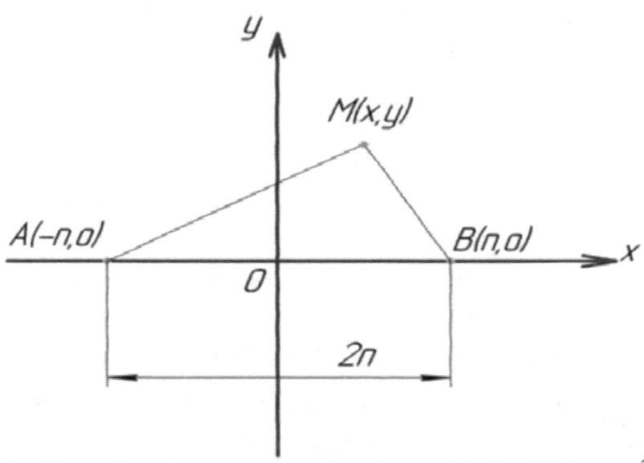

Figure 1. Coordinate system selection: AM · BM = m^2

Let x and y be the coordinates of an arbitrary point M lying on the desired curve (Fig. 1), i.e. M (x, y) - point M with current coordinates. By establishing the relationship between the current coordinates and the constant values characterizing the given curve, we will compose the equation of the curve (geometric location). In this case, we need to establish the relationship between x, y and the constants m2 and n. From the definition of the geometric location it follows (Fig. 1): AM · BM = m^2. Expressing the lengths of the segments AM and BM according to the formula for the distance between two points (the coordinates of points A and B with respect to the coordinate system chosen by us will be A (-n. 0). B (n. 0)), we get (1):

$$\sqrt{(x+n)^2+y^2}\cdot\sqrt{(x-n)^2+y^2}=m^2 \quad (1)$$

This is the equation of this geometrical place. After simplification: freeing from radicals, making possible reductions and elementary transformations, we get: $(x^2+y^2+n^2+2n\cdot x)\cdot(x^2+y^2+n^2\text{-}2n\cdot x)=m^4$, or $(x^2+y^2+n^2)\ 2\ \text{-}4n^2x^2=m^2$, or $(x^2+y^2)\ 2+2\cdot n^2\,(x^2+y^2)\ \text{-}4n^2\cdot x^2=m^4\text{-}n^4$ and finally:

$$(x^2+y^2)^{\,2}\text{-}2\cdot n^2\,(x^2\text{-}y^2)=m^4\text{-}n^4 \quad (2)$$

To build a curve, we examine the found equation. Since x and y (2) are squared in equation, which does not change their sign to the opposite when changing their sign, the curve is symmetric about the coordinate axes. Further, from the geometric definition of the geometric place in question, it is obvious that this curve is a closed curve.

When y = 0, we find the points of intersection with the axis ox:

$x^4\text{-}2\cdot n^2x^2=m^4\text{-}n^4$

When x = 0, we find the points of intersection with the axis oy:

$y^4+2\cdot n^2y^2=m^4\text{-}n^4$

Having built a series of curve points, the coordinates of which can be found, calculating one of the coordinates from the given arbitrary values of the other from the found equation, and connecting them with a smooth line, we get (for m> n) a curve (Figure 2).

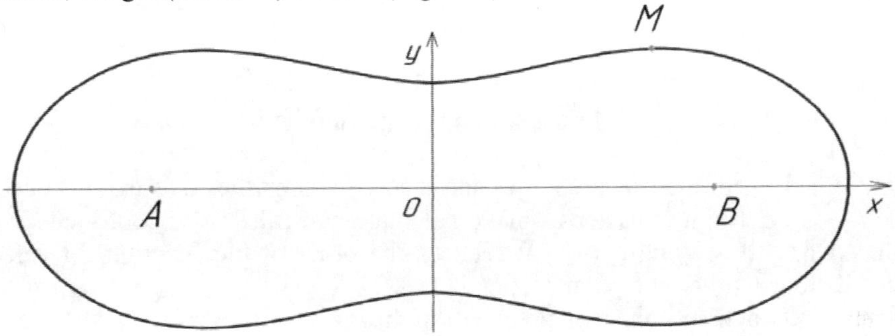

Figure 2. Oval Cassini

a) With the same choice of coordinate axes, the lemniscate equation takes the form (3):

$$(x^2+y^2)^2=2\cdot m^2(x^2\text{-}y^2) \quad (3)$$

b) If the pole is combined with the beginning of the Cartesian coordinate system, and the polar axis is combined with the axis ox, then the lemniscate equation will have the form (4):

$$\rho^2=2m^2\cdot\cos 2\varphi \quad (4)$$

First of all, the position of the coordinate system is established to derive the equation in polar coordinates: we combine the pole with the origin of the selected Cartesian system in the plane, and we align the polar axis with the axis ox. Then we get (Figure 3):

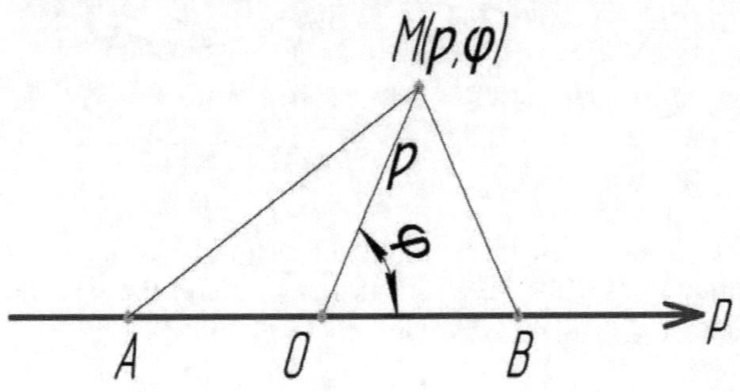

IOP Conf. Series: Materials Science and Engineering **911** (2020) 012011 doi:10.1088/1757-899X/911/1/012011

Figure 3. Polar system selection: AM·BM=m^2

Evaluating the segments AM and BM according to the well-known trigonometry formula ($x = \rho\cos\varphi;\ y = \rho\sin\varphi;\ x^2 + y^2 = \rho^2$), we obtain equation (5):

$$\sqrt{\rho^2 + m^2 + 2m\cdot\rho\cos\varphi}\cdot\sqrt{\rho^2 + m^2 - 2m\cdot\rho\cos\varphi} = m^2 \tag{5}$$

Performing elementary transformations, we obtain:

$(\rho^2 + m^2)^2 - 4m^2\rho^2\cdot\cos^2\varphi = m^4;\ \rho^4 = 4m^2\cdot\rho^2\cdot\cos^2\varphi - 2m^2\cdot\rho^2,$

or $\rho^4 = 2m^2\rho^2\cdot(2\cos^2\varphi - 1)$ and finally (6):

$$\rho^2 = 2m^2\cdot\cos2\varphi \tag{6}$$

This curve is shown in Figure 4.

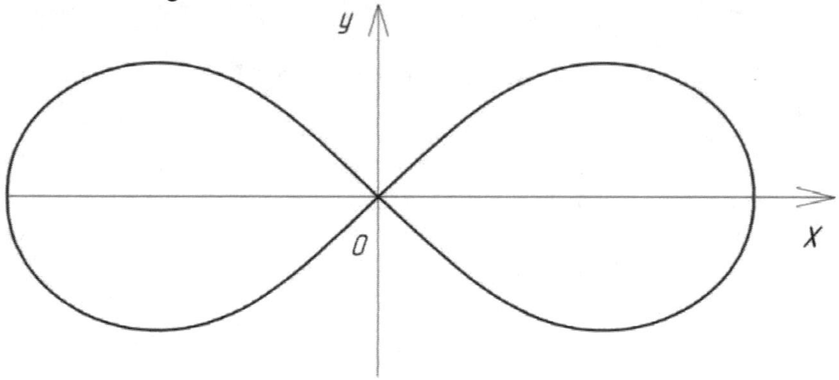

Figure 4. Lemniscata Bernoulli

A comparative analysis of the executed lines and their equations shows us that the accuracy of their calculations is the same, but it is easier to make calculations in the polar coordinate system for the lemniscate, since explicitly setting ρ = f (φ) reduces the order of the function for calculations (the Cartesian system is fourth order, the polar system is second order).

This approach finds a wide range of practical application, starting from a summer residence, a platform in front of a house or school to the architecture of the future, for breaking a flower garden or an exotic arrangement of plants to building castles, windows, roofs in space, these lines form surfaces of interesting cylindrical shape in the Cartesian coordinate system, when the Z-axis takes any value. (Figure 5).

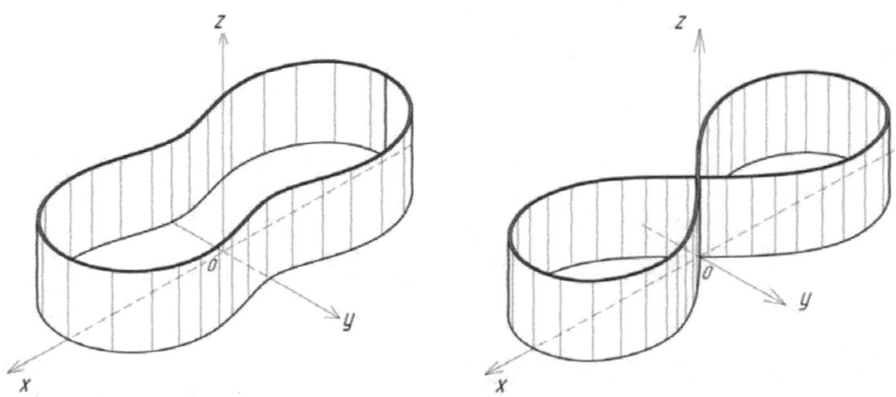

a) b)

Figure 5. Cylindrical surfaces: a) Oval Cassini; b) Lemniscata Bernoulli of straight lines, called rectilinear generators of these surfaces.

The surfaces defined by equations (2) and (3) are cylindrical and called the Cassini and Bernoulli cylinders, respectively. Their generators are parallel to the oz axis, and the guides are the Cassini oval and the lemniscate Bernoulli in the oxy plane.

And with the help of structural hardening by means of straight-line generators (Figure 6), interest in them will increase in the near future, perhaps, they will open up a universal approach to composing equations and surfaces to predetermined geometric parameters for high-order nonlinear functions, with research on the possibility of their hardening [2].

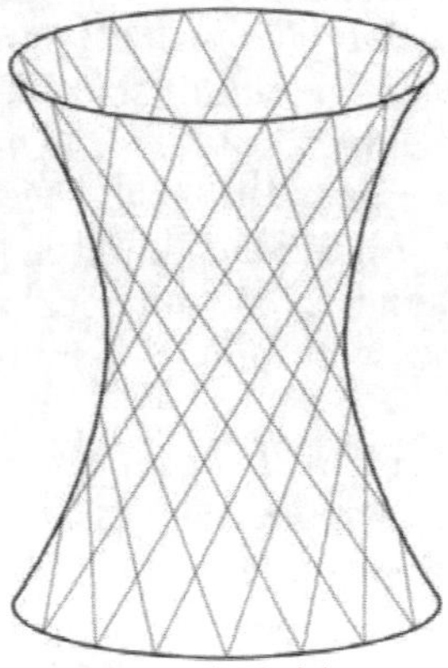

Figure 6. One-sheeted hyperboloid composed of straight lines called rectilinear generators

4. Conclusion

1. You should choose a coordinate system for designing building constructions of a specific geometric location with predetermined properties. (Figure 1, Figure 2).
2. After choosing a coordinate system, you should determine the location of the parameters specified in the project for the given values in these coordinates by designing them on a plane.
3. On the desired line, you should select an arbitrary point with the current coordinates (Figure 1, Figure 2, point M).
4. Having established the relationship of current coordinates with constant values (parameters) according to the project condition, compose the equation: formulas (2), (3), (4).
5. After simplifying the compiled equation, construct a building structure (surface) by examining it for properties of interest.
6. By activating your intuition, your experience will tell you which coordinate system to choose and how to place the given parameters in them in order to simplify the finding of the equation for constructing the construction.
7. If you design the surface and the location of constant values (parameters) in it, the correctness of the selected coordinate system on a plane or in space will make it possible to make a less complex function, taking into account all the relationships between the geometrical location of the points and the parameters under study.

In conclusion, we note that having mastered this algorithm for compiling building structures, according to predetermined conditions, you can design constructions of any degree of complexity for their construction, which is especially in demand in construction equipment. Considering the problems encountered in the design as additional features [3-11], we open the curtain of endless transformations of some forms of lines and surfaces into others, as well as the variety of methods for their hardening, by connecting links of one chain of rectilinear generators. Having created a database of elements of building structures and their connecting links, we come to the possibility of automating the process of designing structures with predetermined properties.

References

[1] Grigorenko N L, Grigorieva É V, Khailov E N and Roi P K 2019 Optimal Control Problems for a Mathematical Model of the Treatment of Psoriasis *Computational Mathematics and Modeling* **30** pp 352–363
[2] Mohapatra J and Mahalik M K 2018 An Initial Value Method for Solving Singularly Perturbed Boundary Value Problems Using Adaptive Grids *Computational Mathematics and Modeling* **29** pp 48–58
[3] Voskoglou M A 2017 Note On The Graphical Representation Of The Derivatives *Physical and Mathematical Education: scientific journal* **2(12)** pp 9-16
[4] Pandey P K 2018 A numerical technique for the solution of general eighth order boundary value problems: a finite difference method *Ural Mathematical Journal* **4 (1)** pp 56-62
[5] Pandey P K 2013 Fourth Order Finite Difference Method for Sixth Order Boundary Value Problems *Computational Mathematics and Mathematical Physics* **53 (1)** pp 57-62
[6] Viswanadham K N S K and Ballem S 2014 Numerical solution of eighth order boundary value problems by Galerkin method with quintic B-splines *International Journal of Computer Applications* **89 (15)** pp 7-13
[7] Reddy S M 2016 Numerical solution of eighth order boundary value problems by Petrov-Galerkin method with quintic B-splines as basic functions and septic B-splines as weight functions *International Journal of Engineering and Computer Science* 5 (09) pp 17894-17901
[8] Jiang Z W 2014 A meshfree method for numerical solution of nonhomogeneous time dependent problems *Abstract Appl. Anal* 978310
[9] Voloshinov D V 2010 *Constructive geometric modeling. Theory, practice, automation: monograph* (Saarbrucken: Lambert Academic Publishing) p 355
[10] Vabishchevich P N, Vasil'ev V I and Vasil'eva M V 2015 Computational identification of the right hand side of a parabolic equation *Comput. Math. Math. Phys.* **55 (9)** pp 1015-1021
[11] Schmitz K S 2018 Chapter 1: Philosophy of Science *Physical Chemistry* pp 183-367

IOP Conf. Series: Materials Science and Engineering **911** (2020) 012012 doi:10.1088/1757-899X/911/1/012012

Assessment of energy efficiency of yielding supports for reinforced concrete beams under dynamic loading

O G Kumpyak[1], Z R Galyautdinov[1] and D R Galyautdinov[1]

[1]Tomsk State University of Architecture and Building, Department of Building materials, 2 Solyanaya Sq., 634003, Tomsk, Russian Federation

E-mail: DaudG@yandex.ru

Abstract. When designing reinforced concrete structures subjected to intense dynamic loading, it is necessary to take into account the simultaneous influence of the thrust reaction, which leads to significant increase in strength and crack resistance, and the influence of yielding supports used to increase the energy intensity of structures. The purpose of this study is to assess feasibility of yielding supports application in thrust structures. The paper presents results of experimental studies of reinforced concrete beams with thrust on yielding supports under short-term dynamic loading. The paper considers the influence of thrust on strength, deformability and crack resistance of reinforced concrete structures under short-term dynamic loading, as well as the joint application of yielding supports and limitation of horizontal displacement of the support contour. The results of the study indicate positive effect of yielding supports application in structures with thrust.

1. Introduction

The factors of emergency impact on building structures can be referred to as: damage and defects in building structures, technical problems with equipment, deviations from design solutions during construction and installation of hazardous industrial facility, high wear of equipment, and insufficient level of new technologies implementation. The occurrence of an emergency is becoming an increasingly urgent problem every day, which leads to significant socio-economic problems: destruction of structure or its part, suspension or complete cessation of all technological processes, injuries and deaths of people providing service for the production process.

When designing buildings and structures, special attention should be paid to enhancement of the methods for calculating and analyzing the stress-strain state of building structures under short-term dynamic emergency loads [1-7].

The paper presents assessment of energy efficiency of yielding supports application [3, 7] in reinforced concrete beams with thrust under intense dynamic loading. As an example, authors provide the analysis of test results of reinforced concrete beam under the action of short-term dynamic loading, as well as reinforced concrete beam with thrust on yielding supports under short-term dynamic loading.

2. Materials and methods

In order to assess energy efficiency of yielding supports and the stress-strain state of reinforced concrete beams with and without thrust under short-term dynamic loading, two design structures were tested: B – beam without thrust on non-movable supports and BS – beam with thrust on yielding supports.

Reinforcement is performed by spatial framework for both B sample and BS sample. As transverse reinforcement, cold-deformed reinforcement Ø5 mm of Bp500 grade was used, installed with a pitch of 50 mm in the support zone and 130 mm in the middle of the span; and as longitudinal reinforcement – hot-rolled reinforcement Ø 6 mm of A240 grade in the compressed zone and Ø10 mm of A500 grade in the extended zone was used. To strengthen the end sections of the beams, authors used grids with a cell of 50 × 50 mm from cold-deformed reinforcement Ø5 mm of Bp500 grade with 7 grids on each side and angles of 100 × 10 mm were installed.

During the test, a set of measuring instruments was placed on the structures (Figure 1): for determining displacements – Waycon RL150 series inductive displacement sensors (Figure 1, pos. 2); for measuring accelerations – accelerometers (DHE 100023) (Figure 1, pos. 3); for determining the input force impact – force-measuring strain gauge DST 4126 (Figure 1, pos. 4); to determine the output force impact – force-measuring strain gauge DST 4126 in the amount of 2 pcs. per each support (Figure 1, item 5).

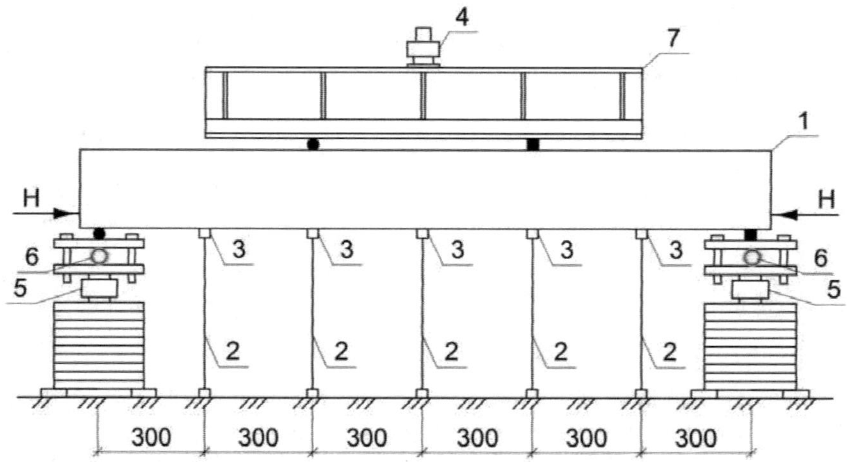

Figure 1. The scheme of placement of measuring instruments while testing of reinforced concrete beams B and BS samples

Visualization of crack formation during short-term dynamic loading is conducted using high-speed camera Photron Fastcam SA-2 with shooting speed of 2500 shots per second.

3. Results

As a result of analysis of displacements of test samples (Figure 2) under intensive dynamic loading, it was found that application of yielding supports and the presence of thrust leads to significant decrease in the structural displacements, which, in turn, leads to decrease in the number of cracks, the depth of their propagation and their opening width, thereby increasing the durability and resilience of structures (Figure 2, Figure 3).

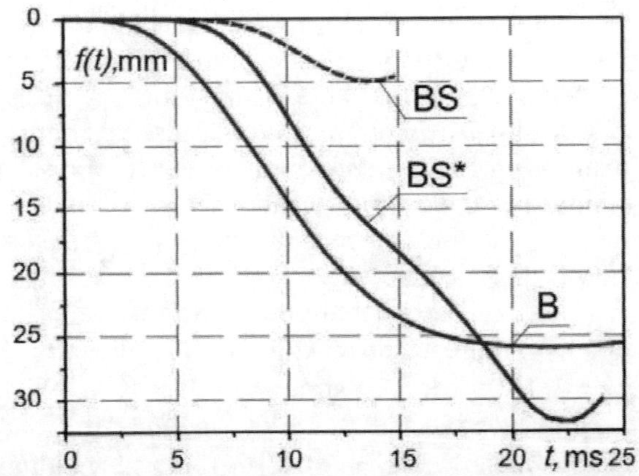

Figure 2. Change in displacements in the middle of the span in time for the samples B and BS (BS* - total displacements of structure and yielding supports)

(a)

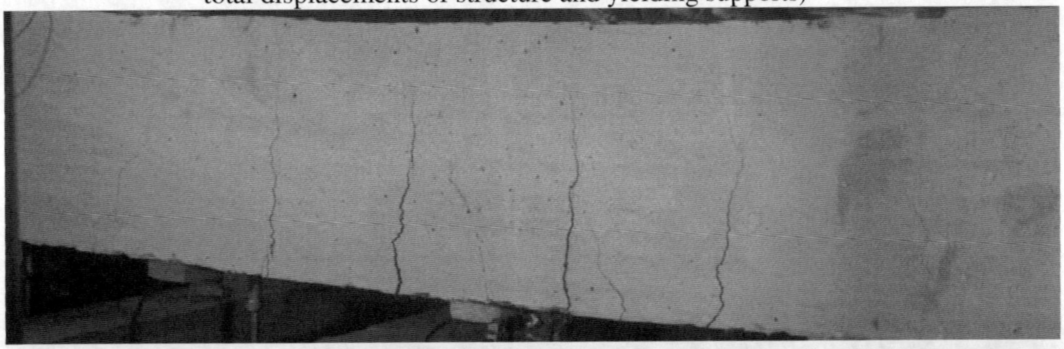

(b)

Figure 3. Crack formation in the ultimate stage for the samples B (a) and BS (b)

One of the ways to evaluate the effectiveness of yielding supports is to build the averaged deformation energy over time for the tested samples.

To obtain the averaged deformation energy over the load action interval, we multiply $F_{def}(t)$ per the averaged displacements of the structure along the length of the element:

$$E_{med}(t) = F_{def}(t) \times f_{med}(t)$$

where $E_{med}(t)$ – is the averaged deformation energy at the interval of load action; $F_{def}(t)$ – is force consumed for the sample deformation; $f_{med}(t)$ – are averaged displacements of the sample along the length at the interval of the action of above-level force impact (the average value of displacements over the time along the length of the sample).

According to the dependence given above the averaged energy diagrams (Figure 4) for the samples B and BS were built.

Analysis of energy distribution diagrams over the time shows that the structure and yielding support are deformed synchronously within the studied time range. For the structures on yielding supports (Figure 4b) up to the time moment of ~ 15 ms the change in the averaged deformation energy of yielding support is insignificant due to the elastic deformations of yielding support. After the transition of support to the plastic stage (time more than 15 ms), a sharp increase in the averaged energy of the beam and yielding support is observed. At the same time, it can be noted that at all stages of deformation, the main part of energy of external impact is absorbed due to deformation of yielding supports (Figure 4, curves 2 and 3). Deformation energy of the beam is significantly lower than the corresponding value for yielding supports at any given time, regardless of the stage of deformation of yielding support (Figure 4, curve 4).

Quantitative estimation can be performed based on the extreme values of deformation energy. As can be seen from energy diagrams (Figure 4) the maximum value of energy $E^{B}_{med} = 1050\,J$ – for the structure of B type and $E^{BS}_{med} = 1150\,J$– for the structure of BS type. Notably, for the sample BS in Figure 4b redistribution of energy is reflected which is accounted for the deformation of structure $E^{BS}_{med,c} = 195\,J$, and energy accounted for the deformation of yielding support$E^{B}_{med,f} = 950\,J$. Therefore, increase in the values of energy consumed for the beam deformation can be characterized by the energy coefficient $k_{E,med} = \frac{E^{B}_{med}}{E^{BS}_{med,c}} = \frac{1050J}{195J} = 5.38$, its physical meaning is that the energy applied to the structure during testing of a sample of BS-2 series was reduced by 5.38 times due to energy absorption of yielding supports, and the energy intensity coefficient of yielding support was $k_{E,flex,b} = \frac{E^{B}_{med,f}}{E^{BS}_{med}} = \frac{950J}{1150J} = 0.826$.

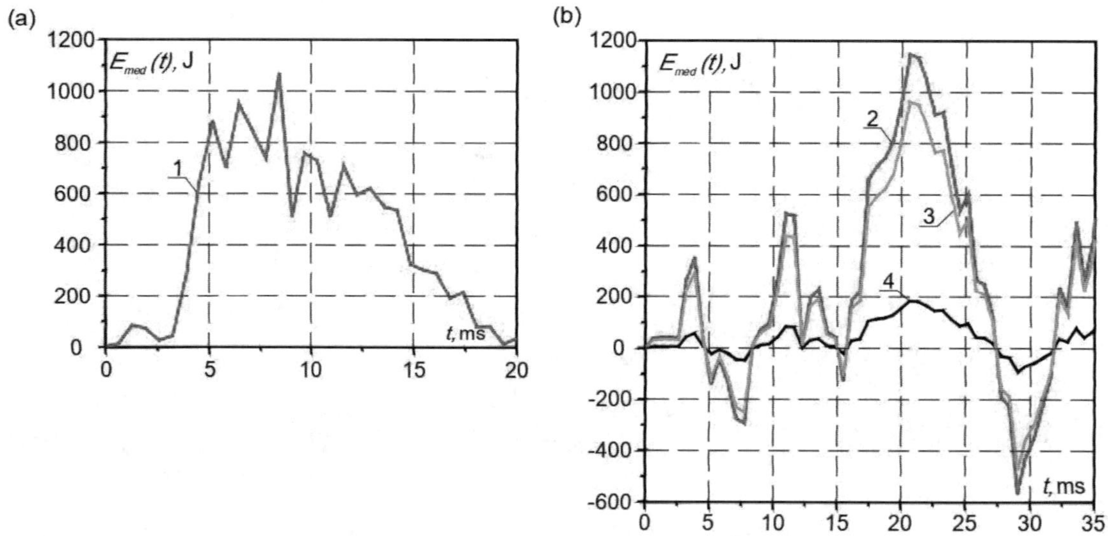

Figure 4. The diagram of averaged deformation energy over time for the samples B (a, diagram 1) and BS (b, diagram 2); the diagram of averaged deformation energy of yielding support over time for BS sample (b, diagram 3); the diagram of averaged deformation energy without taking into account crushing for BS sample (b, diagram 4)

4. Conclusion

The results of the study indicate the positive effect of yielding supports application in structures with thrust under intense dynamic loading. Based on the conducted tests, decrease in the energy consumed on deformation of the structure is observed due to energy redistribution between the structure and yielding support with equal parameters of impact loading. Reinforced concrete beam with thrust on yielding supports is more energy-intensive, which is confirmed by the coefficient $k_{E,flex,b}$ and indicates a clear reduction in energy due to the use of yielding supports.

5. Acknowledgements

The reported study was funded by RFBR, project number № 19-31-90097.

References

[1] Vinogradova T 1977 *The influence of thrust on the operation of reinforced concrete beams under short- term dynamic loading* PhD thesis (Moscow) p. 155 (in Russian).

[2] Korsun V, Nedorezov V and Vinogradova T 2018 The influence of thrust on the load-bearing capacity of reinforced concrete beams *Loleit's Readings – 150* (Moscow) pp 180-186

[3] Kumpiak O and Galiautdinov Z 2012 Tomsk Calculation of reinforced concrete beams on yielding supports under short-term dynamic loading *Vestnik TSUAB* **2** pp 107-111

[4] Plevkov V, Sarkisov D and Baldin S 2017 Analysis of parameters of oscillating processes in load- bearing frame of industrial building *Building and reconstruction* **4** pp 47-56

[5] Odnokopylov G and Sarkisov D 2017 Evaluation of parameters of non-destructive loading under shock-wave loading for the major building structures of oil and gas industry *Bulletin of the Tomsk Polytechnic University. Geo Assets Engineering* **328** pp 85-95

[6] Odnokopylov G, Sarkisov D and Butuzov E 2018 The evaluation of the durability degree of major structures under shock-wave loading *Bulletin of the Tomsk Polytechnic University. Geo Assets Engineering* **329** pp 122–135

[7] Kumpyak O, Galyautdinov Z and Kokorin D 2016 Strength of concrete structures under dynamic loading *AIP Conference Proceedings*

IOP Conf. Series: Materials Science and Engineering **911** (2020) 012013 doi:10.1088/1757-899X/911/1/012013

Simulation of the strain diagram of a slag-concrete element subject to bending

N Chernousov[1], B Bondarev[2] and V Sturova[3]
[1]General Director NTO EXPERT, 9 Kommunalnaya Square, Lipetsk, Russian Federation
[2]Doctor of Technical Sciences, Professor of the Department of Building Materials Science and Road Technologies, Lipetsk State Technical University, 30 Moskovskaya Street, Lipetsk, Russian Federation
[3]Master, post-graduate student of the Department of Building Materials Science and Road Technologies, Lipetsk State Technical University, 30 Moskovskaya Street, Lipetsk, Russian Federation

E-mail: v-livenceva@mail.ru

Abstract. In the modern world of information technology personal computers considerably dominate our everyday life. All real, field experiments and tryouts replace computer-aided simulations as they often save a lot of time. It is much easier to make the multiple calculations related inter alia to reinforced concrete structures using strain diagrams for the concrete and re-bars. This is the method ensuring the results similar to the field tests and the only challenge is that many equation parameters have to be reduced. Therefore, the objective of the work is to propose a simplified mathematical model for describing the strain diagram of a bended slag-concrete element based on air-cooled slag (ACS) and the aggregate made of sieves (0...5 mm fraction) from crashed ACS originating from NLMK's Blast Furnace Operations, while implementation of the model helps to avoid complicated equilibrium experiments. As a result of the research and mathematical simulation of the bended element diagram a model capable of reflecting the sample behavior at any loading stage using maximum load and initial modulus of elasticity was proposed.

1. Introduction

Due to intensive development of IT-equipment we face a problem of scheming an automated method for computation and design of building structures and creating such mechanical models of concrete and reinforced concrete which are general enough to approximate the state-of-the art strength models and theories of deformable solid body mechanics and closely related thereto [1].

No development of modern enhanced computation methods for reinforced concrete structures in terms of strength and far less deformability is feasible without considering nonlinear properties of concretes: heavy-weight high-strength concretes [2], light-weight concretes with artificial porous aggregates [3] as well as concretes using by-products of different industries and local aggregates [4,5].

It is noted that structural damage and deformability is strongly impacted by the features of the aggregate. For dense (heavy-weight) concretes four compression failure stages are specified [2]: 1) initial loading stage having intensity up to about (0.1-0.15) R_b with characteristic development of the concrete permanent deformation, low modulus of elasticity; 2) the stage of the sample volume compaction and reduction when loading with intensity up to about (0.15-0.3)R_b.; 3) the stage of the

IOP Conf. Series: Materials Science and Engineering **911** (2020) 012013 doi:10.1088/1757-899X/911/1/012013

material decompaction and disintegration (the first reference point) - characterizes the threshold of the concrete elastic behavior at the moment when first micro cracks are generated in the concrete structure when loading with intensity up to about (0.3-0.6) R_b; 4) extreme stage of the concrete equilibrium condition effecting the cracks (beyond the boundaries of which the proportion of stresses and deformation is upset - characterizes the threshold (the second reference point) when loading with intensity up to about (0.5-0.85) R_b.

Experiments show [3] that for light-weight concretes not four but three areas of stressed conditions are revealed. Such statement is supported by higher deformability of porous vs. compact aggregate by its ideal bond with mortar fraction and shrinkage-induced 3D reduction. Hence, for light-weight concretes based on artificial porous aggregates [3], as well as concretes based on air-cooled slag (ACS) and the aggregate made of sieves (0...5 mm fraction) from crashed ACS originated from NLMK's Blast Furnace Operations [5] the level fluctuation for the first parametrical point is possible within (0.3-0.7) R_b (i.e. higher than for heavy-weight concretes), and for the second parametrical point this level comes up to (0.1-1.0)R_b with concrete strength approximately identical.

When developing a model of the structure real behavior it is necessary to take into account an important feature of the experimental diagrams of the concrete compression [2], characterized by the fact that at the time of initial load application to the sample, which has never been loaded before, significant residual strain is revealed characterized by low values of the modulus of deformation. Normally, when testing concrete samples, these deformations are not noted while they reveal themselves during the sample centering for the next test. Nevertheless, these deformations impact the real behavior of the structure especially the one non identified statically.

For analytical description of the entire concrete deformation model the third reference point should be taken into account (the threshold condition of the concrete) beyond which significant crack formation and concrete failure take place. The second and third reference points in the ultimate design behavior are supposed to be determined simultaneously as the meeting point of the respectively rising and down-coming curves of the real diagram with horizontal section [6].

The stress-deformation method for computation of reinforced concrete structures using strain diagrams of concrete and re-bars has recently attained high priority as ensuring high level of reliability in assessing the strength and deformational properties [7-14].

Along with deformational models the engineering practice has a long experience of using conventional computation method based on the ultimate behavior approach.

Research shows that for reinforced concrete elements with regular shapes (rectangular, T-beams, I-beams) the strength computations can be made using simplified diagrams of the material behavior [16-18].

Based on comparative analysis [19] the conclusion has been made that the strength calculations obtained using simplified stress-deformation model and those according to SP 63.13330.2012 «Concrete and reinforced concrete structures», differ greatly, though, the curve and respectively the bends in the middle of the span calculated using the stress-deformation model is a sequence higher than according to code of practice (SP), therefore, it is necessary to limit the norms for limit and threshold values of stress deformation in the diagrams of the concrete deformation induced by axial compression and the re-bar deformation induced by axial tension.

2. Materials and methods

Samples of slag-concrete, bending test.

In order to fully reflect the real physical processes of damage accumulation in the material during mechanical tests it is recommended to use fully stable strain diagram (FSSD) obtained during bending test [20].

It is known that during bending test using the samples of elastoplastic material (e.g. a slag-concrete element) inelastic bending can sometimes take place, i.e. the material fails to follow Hooke Law. The simplest case of inelastic bending can be plastic hinge taking place with elastic-ideally-plastic material. Such material follows Hooke Law until the tension value reaches the yield limit and then plastic deformations under constant tension [21] start developing.

As an example, we consider the behavior of a rectangular-shape freely supported beam made of elastic-ideally-plastic material under concentrated force ***P*** applied right in the middle (Figure 1,a).

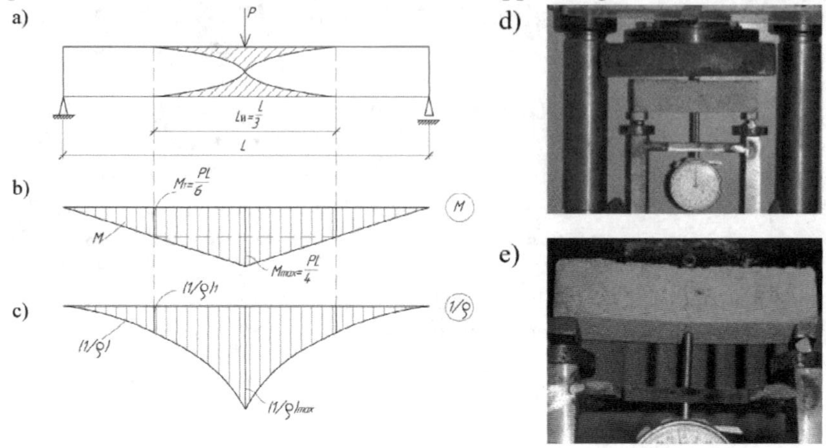

Figure 1. Plastic zone and plastic hinge: a) appearance of plastic zone and generation of plastic hinge; b) bending-moment diagram; c) curve diagram; d) general view of three-point bending test of the sample; e) pre-fracture of the sample during three-point bending

Bending-moment diagram represents a triangle and max. momentum M_{max} equals to $PL/4$ (Figure 1,b). When max. momentum value is higher than M_T $=PL/6=0.67M_{max}$, but less than ultimate momentum M_{ult} (M_{ult} = M_{max} = $P_{max}L/4$), then limited plastic yield will take place in the central part of the beam, while the plastic zone length for a rectangular shape will be L_T=L/3 (Figure 1,a).

Figure 1,c shows the curve distribution (1/ρ), depending on the bending moment *M*, modulus of elasticity *E* and second area moment *I*, and equals to *M/(EI)*. Bends in the middle of the span were measured using dial test indicator accurate to 0.001 mm (Figure 1,d). The curve grows following the linear law in the section from the beam end to the beginning of the plastic zone, where the curve equals to $(1/\rho)_T$. Then the curve growth rate increases and in the middle of the beam the maximum value $(1/\rho)_{max}$ is obtained. The maximum curve value stays final until elastic zone is maintained in the middle of the beam. When the load increases even more and the value of maximum bending moment approaches the ultimate moment value M_{ult}, the plastic zones in the middle of the beam start quickly developing inside towards the neutral axis. Finally, when M_{max} reaches M_{ult} value, the cross section in the middle of the beam completely turns into plastic (Figure 1). The curve in the middle of the beam becomes extremely high, the bend in the middle of the span quickly increases which is measured during the experiments using a dial test indicator accurate to 0.001 mm (Figure 1,d) and unlimited plastic yield appears. The maximum momentum value cannot grow any more while the load reaches the peak value. The beam is weakened by extreme turns appearing in the middle cross section while the two parts of the beam remain relatively rigid. Thus, the beam behaves like two rigid rods, connected by a plastic hinge allowing these two rods rotate against each other influenced by constant momentum M_{ult} (Figure 1,e).

Hence, in case of inelastic bend, it can be assumed that the material has followed Hooke Law at the momentum value of M = 0.9 M_T = $0.6P_{max}L/4$ = $0.6M_{max}$.

The same assumption can be made for the trial beams used for four-point-bending test (Figure 2).

IOP Conf. Series: Materials Science and Engineering **911** (2020) 012013 doi:10.1088/1757-899X/911/1/012013

a)

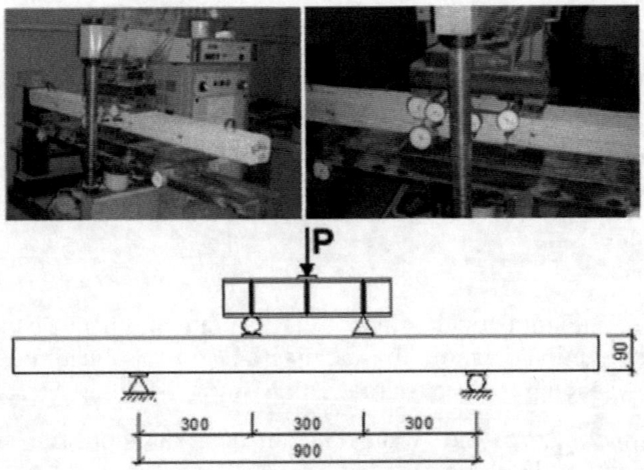

b)

Figure 2. Four-point-bending test of a beam: a) test general view; b) load application chart

The value of the initial modulus of elasticity can be obtained from the compression test or by measuring US speed in the material, or directly measured during bending [18,22,23].

Determine the value of the initial modulus of elasticity during bending at M= $0.6M_{max}$ and f_y= $0.6f_{max}$. It is known that the bend in the middle of the span at the elastic stage of the beam behavior equals to [21]:

- for three-point bending
- for four-point bending

Then from formulas (1) and (2) determine the value of the initial modulus of elasticity:

- for three-point bending
- for four-point bending

In formulas (3) and (4) k_M= L^2/I- single yield during bending, mm^{-2}.

During computation of reinforced concrete structures by stress deformations based on non-linear deformation model (NLDM) we should use the modulus of deformation (transversal or tangential) which, as opposed to the initial modulus of elasticity (E_b is a constant value), changes with the increase of the load value and exposure time, and tends to zero when they decrease while deformations achieve their extreme values [1,6,7-13,18].

With multiple dependencies proposed the problem here is to minimize the equation parameters.

Analyzing behavior of the real structures and models made of heavy-weight and light-weight concrete [2,3], as well as structures based on slag-pumice-concrete [5,18,23], in order to make mathematical description of the dependency between the applied load (M) and the bend (f) during bending we assume the following parameters as characteristic parameters of the diagram "*M – f*" (Figure 3): angle α of the rising curve of the diagram to abscissa axis ($tg\ \alpha = 0.6M_c/f_1$), point 1 at the end of the straight-line portion of the rising curve, point 2 the peak of the deformation curve, point 3 bends of the down-coming curve of the diagram and point 4 the sample defragmentation.

Analysis of the common strain diagrams for concrete and re-bars [1,6-14,18] provides for describing the diagram using the following expression:

where *A*, *B* and *C* – parameters of the deformation curve

Parameters *A*, *B* and *C* of the equation (5) shall be found based on physical and geometrical assumptions [18], as well as on the obtained test data.

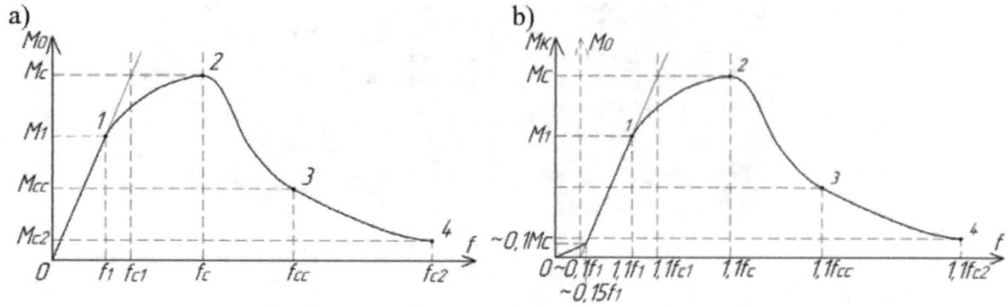

Figure 3. Parametrical points of the "bending moment - bend" diagram
a) assumption diagram; b) real diagram (first application of the load)
Mstr – bending moment (structure); Msmp – bending moment (sample)

To validate the simulation of the strain diagram of a slag-concrete element subject to bending experimental strain diagrams [18,24]were used.

Compound №1 of B20 class concrete, based on ACS (sample BSh-1) from «Zhelezobeton Plant» (Lipetsk) for production of self-supporting structures (in particular, bar-type lintels of 1.038.1-1 series) and experimental compound №2 of fine-grain concrete matrix (sample OA-0-0-6) are shown in Table 1.

Table 1. Compounds of concrete based on ACS and sieves from crashing

No. Item No.	Component	Component consumption, kg/m ^{3}	
		Compound №1	Compound №2
1	Portland-blastfurnace cement M400	328	531
2	Air-cooled slag (5-20 mm fraction)	1030	-
3	Sieves from crashed ACS (0-5 mm fraction)	-	1423
4	Sand (pit-run)	700	-
5	"Universal P-2" additive	19	30.6
6	Water	170	224

Table 2 shows geometrical characteristics and main physical and mechanical properties of concrete of the beam samples at the time of testing.

Table 2. Sample characteristics

Sample grade (age of concrete)	Loading chart	Geometrical characteristics					Initial modulus of elasticity E_b, MPax10^{-3}	
		b, mm	h, mm	L, mm	I, mm^4	k_M, mm^{-2}	During compression	During bending acc. to formulas (3) and (4)
OA-0-0-6 (714 days)	Figure 1	40	40	150	213333.33	0.105	33.6	34.54
BSh-1 (912 days)	Figure 2	120	90	900	72900000	0.111	22.7	22.94

Figure 4 shows experimental strain diagrams of the samples. Parameters *A*, *B* and *C* of the equation (5), found based on physical and geometrical assumptions [18], are shown in Table 3. Parametrical points of the diagram (Figure 2) based on the obtained test data are given in Table 4.

Table 3. Stress deformation values

Sample	A, N/mm	B	C,1/mm
OA-0-0-6	1455422.2	0.7066	-7.8508
BSh-1	71506-90.9	0.3742	-0.6803

Table 4. Values of the strain diagram parametrical points

Parametrical points	Point «1»		Point «2»		Point «3»		Point «4»	
	$0.6\,M_c$, Nmm	$0.6f_{c1}$, mm	M_c, Nmm	f_c, mm	M_{cc}, Nmm	$1.5f_c$, mm	M_{c2}, Nmm	$2f_{c2}$, mm
OA-0-0-6 (experiment)	120300		130988		108863		83400	
OA-0-0-6 acc.to formula (5)	118376	0.05	130998	0.09	120888	0.14	1055471	0.18
BSh-1 (experiment)	354000		393300		-		-	
BSh-1 acc.to formula (5)	380520	0.35	393315	0.55	377445	0.85	362201	1.0

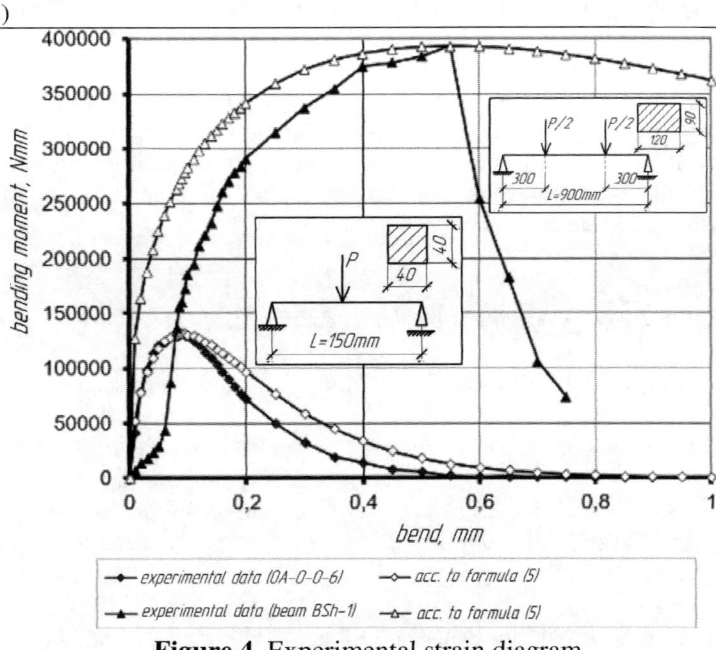

Figure 4. Experimental strain diagram

3. Results

As is seen from Table 4 and Figure 4 the down-coming area shows deviation between the experimental and theoretical (5) values of the applied load, therefore, the formula (5) should be adjusted taking into account the dynamic movement of the main crack. It is also noted that for beam BSh-1 the formula (5) does not take into account the real behavior of the structure at the initial loading stage ($M=0.1M_{max}$).

In order to eliminate the revealed deviations and to simplify the mathematical model for describing the material strain diagram during bending via single parameter M_c (max. load value) providing that the modulus of elasticity is known, we can express the diagram description as follows:

The method to identify A and B parameters is given in [18].

Below the expression, please, find «M-f» dependencies:

Using formulas (7) - (12) and taking into account the experimental values in Figure 5 a fully stable diagram "M – f" can be plotted.

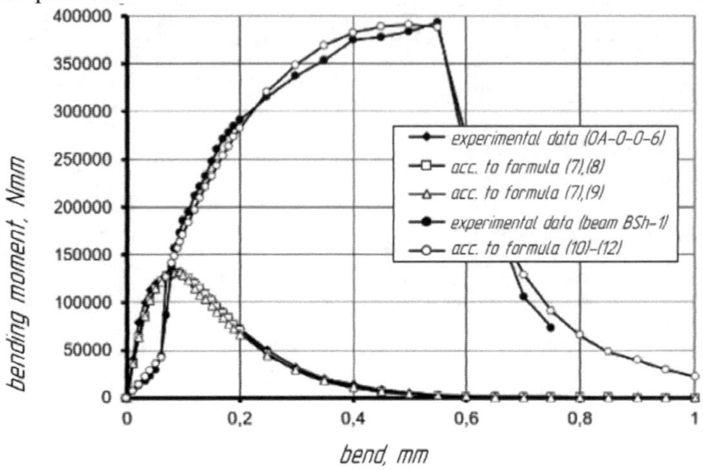

Figure 5. Strain diagram "M – f", plotted acc. to formulas (7)-(12), vs. experimental values

4. Conclusion

The proposed mathematical model for describing the fully stable strain diagram of a slag-concrete element subject to bending using a single experimental parameter (max. load M_c) and known initial modulus of elasticity of the material allows to avoid complicated equilibrium experiments related to constant recording of the bends at each loading stage and is capable of simulating the material behavior at each loading stage up to the material failure.

References

[1] Karpenko N 1996 *General models of reinforced concrete mechanics* (Moscow: Stroyizdat) p 416.
[2] Berg O, Shcherbakov E and Pisanko G 1971 *High strength concrete* (Moscow: Stroyizdat) p 208.
[3] Ivanov I 1993 *Lightweight concretes on artificial porous aggregates* (Moscow: Stroyizdat) p 182.
[4] Ufimtsev V and Korobeinikov L 2014 *Concrete technologies* **6** pp 50-53.
[5] Chernousov N and Pantelkin I 1998 *Reinforced concrete structures using dispersed-reinforced slag-pum-concrete: monograph* (Publishing house of the Association of building universities) p 230.
[6] Maiorov V and Ratsirinivu de Roussel Gilbert 2000 *Bulletin of the Peoples' Friendship University of Russia. Series: Engineering Research* **3** pp 97-102.
[7] Karpenko N and Radaykin O 2012 *Construction and reconstruction. Orel: Publishing house of OSTU* **2 (40)** pp 11-18.
[8] Karpenko N and Radaykin O 2012 *Construction and reconstruction. Orel: Publishing house of OGTU* **3(41)** pp 10-17.
[9] Karpenko N, Sokolov B and Radaykin O 2016 *Izvestiya vuzov. Building* **3** pp 5-12.
[10] Panfilov D, Pishchulev A and Gimadetdinov K 2014 *Industrial and civil construction* **3** pp 80-83.
[11] Varlamov A, Shishlonov E, Tkach E, Shumilin M and Goncharov D 2016 *Academy* **2(5)** pp 7-16.
[12] Radaikin O 2017 *Bulletin of civil engineers* **6(65)** pp 71-78.
[13] Radaikin O 2019 *Vestnik BGTU* **10** pp 29-39.
[14] Fedorov V, Shavykina V, Fedorov M and Yusupova E 2017 *Construction and reconstruction*

4(72) pp 80-85.
[15] SP 63.13330.2012 Concrete and reinforced concrete structures. Basic provisions. Updated edition of SNiP 52-01-2003. Moscow 2015.
[16] Shah S and Jehu R 1986 Strain rate effects an mode crack propagation in Concrete (Fract. Toughness and Fract. Energy: Coner. Proc. Conf. Lensaune. Oct. 1-3. 1985. Amsterdam e. a.) pp 453-465.
[17] Jeng Y and Shah S 1985 *J. Eng. Mech.* **10** pp 1227-1241.
[18] Chernousov N and Sturova V 2020 *Modern science-intensive technologies* **3** pp 92-96.
[19] Eryshev V 2018 *Systems. Methods. Technology* **1(37)** pp 79-84.
[20] GOST 29167-91 "Concrete. Methods for determining the characteristics of crack resistance (fracture toughness) under static loading."
[21] Pisarenko G, Yakovlev A and Matveev V 1988 *Handbook on the strength of materials* Editor-in-chief Pisarenko G.S.-2nd ed., Revised and supplemented (Kiev: Nauk.dumka) pp 736.
[22] GOST 24452-80. "Concrete. Methods for determining prismatic strength, modulus of elasticity and Poisson's ratio.
[23] Bondarev B, Chernousov N, Chernousov R and Sturova V 2019 *Colloquium-journal* **15-1(39)** pp 4-6.
[24] Chernousov N and Chernousov R 2010 *Concrete and reinforced concrete* **4** pp 7-11.

Vacuum filtration on a thin vertical filtering baffle

E Yu Kurochkin[1] and N Yu Kurochkina[2]

[1]Immanuel Kant Baltic Federal University, 14 Nevskiy St, Kaliningrad, 236016, Russian Federation
[2]Kaliningrad Branch of the Accredited Private Educational Institution of Higher Education "Moscow University of Finance and Law MUFL", 16-1 Denisova St., 236048, Kaliningrad, Russian Federation

E-mail: viv653521@mail.ru

Abstract. The article considers the design issue of vacuum filtering apparatuses from the point of view of minimizing filtering costs, provided that the maximum performance of vacuum filtration apparatuses is ensured. In the paper, the Darcy equation was considered. Based on it, the equations of filtration apparatus productivity were determined by filtrate and by dry sediment. It was determined that the maximum productivity of both filtrate and dry sludge in vacuum filtration apparatuses is observed with a maximum difference in free surfaces on the vertical filter baffle.

1. Introduction

Various devices and mechanisms are used to compact and dehydrate sludge in the practice of water treatment. Some devices work by using the force of gravity (devices with bag filter elements [1]). More devices create pressure higher (for example, filter presses) or lower (for example, vacuum filters) than atmospheric pressure, increasing the productivity of the apparatus [2].

Analysis of scientific literature has shown that despite the variety of applied designs of vacuum filters with different geometric shapes and sizes of filter surfaces, there are no recommendations regulating the calculation of a vacuum filtration installation with vertical filtering baffles.

For example, work [3] considers the process of vacuum filtration and the choice of the optimal filtering surface, taking into account the capital costs of filtering equipment. But when determining the volume of the filtrate for one minute, it was not taken into account that in the process of filtration, the filtration rate (and hence the volume of the filtrate per unit area) does not change linearly, but rather according to the logarithmic dependence [4].

2. Materials and methods

The mathematical description of the filtration process through a horizontal baffle is known [5]. Simulation of the operation of technological equipment is performed according to well-known methods [2, 6]. Consider the process of vacuum filtration through a vertical baffle, the hydrostatic pressure on which increases from top to bottom (Figure 1), while it is different on the free surfaces of the liquid on different sides of the filter baffle. Theoretical and experimental studies of the filtration process made it possible to derive some filtration equations for vertical filter surfaces [7] operating at pressures above atmospheric pressure.

A mathematical model of vacuum filtration of incompressible sediment is considered. The following assumptions are applied:

- the initial suspension does not stratify;
- sediment does not slide along the partition;
- the process takes place at constant pressures P1 and P2;
- the resistance of the filtration baffle is low compared to the resistance of the sediment;
- a flat rectangular vertical filtering baffle is considered (Figure 1).

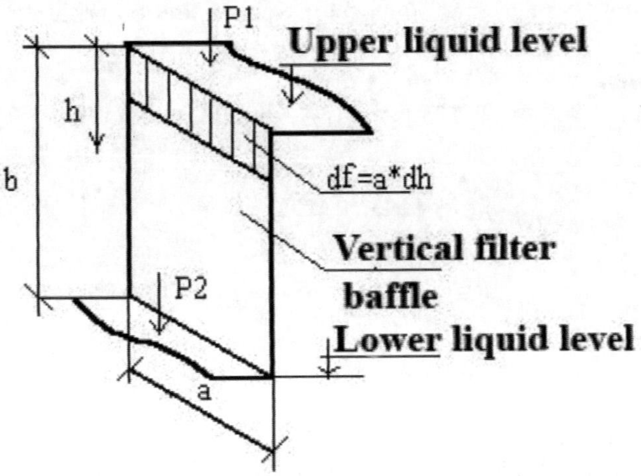

Figure 1. Design scheme of filtration through a vertical baffle.

In this case, we will use the basic filtering equation (Darcy's equation):

$$\frac{dV}{Fdt} = \frac{\Delta P}{\mu r_0 x_0 \dfrac{V}{F}} \qquad (1)$$

where V – filtrate volume, m^3;
F – filtration surface, m^2;
t – filtration duration, s;
ΔP – pressure difference across the filtering partition, Pa;
μ – viscosity of the filtrate, Pa·s;
r_0 — volume resistivity of the sediment, m/kg;
x_0 – proportionality factor in the equation
$h_{oc} = x_0 \cdot (V/F)$,
where h_{oc} is the thickness of the sediment layer on the filtering partition, m.

Filtration constants r_0 and x_0 should be determined experimentally for specific sediment by laboratory experiments [8].

By integrating equation (1) in the range from 0 to V and from 0 to t, as well as performing transformations, we obtain dependence for determining the volume of the resulting filtrate (5):

$$VdV = [\Delta P F^2 / (\mu r_0 x_0)]dt \qquad (2)$$

$$\int_0^V VdV = \int_0^t \frac{\Delta P F^2}{\mu r_0 x_0} dt \qquad (3)$$

$$\frac{V^2}{2} = \frac{\Delta PF^2}{\mu r_0 x_0} t \qquad (4)$$

$$V = \sqrt{2\frac{\Delta PF^2}{\mu r_0 x_0} t} \qquad (5)$$

The value of pressure ΔP under the influence of which filtration is carried out is the total value of the hydrostatic pressure ρgh and the pressure difference Δp on both sides of the filtering baffle on the free surfaces of the liquid, i.e. $\Delta P = 0.5\rho gh + \Delta p$.

For an infinitesimal increase in the area df of the filtering baffle, the increase in the filtrate volume will be expressed:

$$dV = \sqrt{\frac{2(0{,}5\rho gh + \Delta p)\cdot t}{\mu r_0 x_0}} \cdot df \qquad (6)$$

$$df = a \cdot dh \qquad (7)$$

where a – width of the filtering baffle, m;
h – coordinate, for the top of the partition $h = 0$, for the bottom of the partition $h = b$;
ρ – density of the suspension, kg/m^3;
g – acceleration of gravity, m/s^2.

Since the value $[2/(\mu r_0 x_0)]^{0{,}5}$ for the incompressible sediment is constant, we denote it by the symbol A_1.

Hence:

$$dV = A_1\sqrt{\Delta P t}a \cdot dh = A_1 a\sqrt{t}\sqrt{(0{,}5\rho gh + \Delta p)} \cdot dh \qquad (8)$$

By integrating and simplifying equation (8), we obtain equation (11), which allows us to determine the filtrate performance of the vacuum filter:

$$\int_0^V dV = a\sqrt{t}A_1\int_0^h \sqrt{(\Delta p + 0{,}5\rho gh)}dh \qquad (9)$$

$$V = a\sqrt{t}A_1\left[\frac{1}{0{,}5\rho g}\int_0^h \sqrt{(\Delta p + 0{,}5\rho gh)}d(\Delta p + 0{,}5\rho gh)\right] = a\sqrt{t}A_1\left[\frac{1}{0{,}5\rho g}\left(\frac{(\Delta p + 0{,}5\rho gh)^{3/2}}{3/2}\right)\right]_0^h \qquad (10)$$

$$V = \frac{A_1 a\sqrt{t}}{0{,}75\rho g}\left[\sqrt{(\Delta p + 0{,}5\rho gh)^3}\right] \qquad (11)$$

Almost any sediment is compressible. In this case, the dependence of the resistivity of the sediment on pressure can be expressed by the empirical Lewis equation:

$$r_0 = r_1 * \Delta P^s, \qquad (12)$$

where r_1 and s are empirical constants determined empirically for each sediment.

For a real compressible sludge, after appropriate transformations and integration, we get:

$$V=\frac{A_2\sqrt{ta}}{0{,}25\rho g}\left[\frac{(\Delta p+0{,}5\rho gh)^{\frac{3-s}{2}}}{3-s}\right] \tag{13}$$

where

$$A_2=\sqrt{\frac{2}{\mu_0 r_1 x_0}} \tag{14}$$

3. Results

Equations (11) and (13) show that if the filtration process takes place without the use of external pressure Δp, then the specific productivity is proportional to the hydrostatic pressure in the power *(3-s)/2*. In this case, flat elements of minimum thickness, vertically lowered down, have the highest specific productivity with an equal filtration area, since with an equal area of all types of surfaces at flat filtering partitions, the value of *h* is maximum. Thus, the spherical shape, which is quite common in the process of vertical filtration, is the least effective.

As the thickness of the sediment layer on the surface of the filtering baffle increases, the overall resistance to filtration increases. If we assume that at some point in time, for technological reasons, the filtration process stopped, and at this moment on the filter baffle there is a sediment with a volume separated from the suspension with an efficiency of *E*, (unit fraction), having a moisture content W_{sed}, then, in this case, the dry sludge mass will be:

$$m_{oc}=\rho V(1-W_{sed})E \tag{15}$$

Specific productivity G, $[kg/(m^2\cdot h)]$, of the filtering partition on dry matter can be determined by the equation (16):

$$G=m_{oc}/(Ft) \tag{16}$$

Taking into account equation (15), we get:

$$G=\rho V(1-W_{sed})E/(Ft) \tag{17}$$

After substitution of the filtrate volume *V* from equation (11) and simplification, we obtain:

$$G=\frac{A_1\left[(\Delta p+0{,}5\rho gh)^{1{,}5}\right](1-Wsed)E}{0{,}75gh\sqrt{t}} \tag{18}$$

4. Conclusion

It can be summarized that in order to ensure maximum performance of vacuum filtration devices, it is necessary:

- Make the most of the filter cloth for filtering the suspension, and for this the apparatus must be immersed in the suspension, and, if possible, only the unit for regenerating the filter material must remain above the liquid surface;
- For maximum utilization of hydrostatic pressure, the filter housing must be submerged to the maximum permissible depth.

References

[1] Khasanov I Yu, Zapirov R A, Bakiev A V and Gareev M M 1997 *Strainer filter* Patent of Russian Federation No. 2088302. Publ. 27.08.1997

[2] Turovskiy I S 2008 *Osadki stochnykh vod. Obezvozhivaniye i obezzarazhivaniye* (Moscow: DeLi Print) p 376

[3] Sadigova N S 2018 Selection of optimal conditions of filtering processes of suspensions gypsum and of magnesium hydroxide *In the proceedings: Breakthrough scientific research:*

problems, patterns, prospects. Collection of articles of the XI International Scientific and Practical Conference (Penza: Nauka i Prosveschenie) p 64-67

[4] Dzubo V V, Kurochkin E Yu and Butovskiy M E 2008 *Pit'evaya voda* **5** p 2-9

[5] Zhuzhikov V A 1980 *Fil'trovaniye: teoriya i praktika razdeleniya suspenziy* Ed. 4th, rev. and additional (Moscow: Chimiya) p 400

[6] Tarleton E S and Wakeman R J 2007 *Solid/liquid separation: equipment selection and process design* 1st ed. (Amsterdam; Boston: Butterworth-Heinemann) p 448

[7] Marchenko Yu G, Galich R A and Smirnov N S 1995 *Water supply and sanitery technique* **12** p 13-15

[8] Aksenov V I, Kuznetsov I A and Aromanov Yu L 1974 *Water supply and sanitery technique* **1** p 13-16

Designing new sustainable anti-landslide systems under constrained city conditions

T P Kasharina[1], D V Kasharin[1] and D A Polyakov[1]

[1]Federal State Budget Educational Institution of Higher Education, Platov South-Russian State Polytechnic University (NPI), 132 Prosveshcheniya St., 346428, Novocherkassk, Russian Federation

E-mail: kasharina_tp@mail.ru, dendvk1@mail.ru, dan.polyackov2011@yandex.ru

Abstract. This article discusses the development of technical solutions for anti-erosion and landslide impact designs in the design and construction of buildings and structures in cramped urban conditions. They are created from composite nanomaterials. Many foreign and domestic scientists are engaged in such structures from soil-reinforced and soil-filled elements. We also examined the methods of calculating their justification and the results of experimental studies.

1. Measures to eliminate landslide phenomena

Currently, a number of environmental and social problems arise during construction in cramped urban conditions associated with natural and technological impacts, including erosion-landslide phenomena. To address this issue, the authors propose technical solutions and measures to preserve the sustainable development of urban construction. Table 1 below presents measures for the elimination of erosion-landslide phenomena in urban development [1-3].

Table 1. Measures to eliminate erosion-landslide phenomena

Types of changes	The causes of the formation of erosion-landslide phenomena	Suggested activities and solutions
1	2	3
Natural	Change in stress state of the soil Ground and surface water, weathering	Creation of underground structures at the base, loading at the site of the expected bulging of earth masses, horizontal and vertical drainage; inclined slot, microplaning, trays, tracks, etc.
Man-made	The combination of a number of active reasons	Mechanical resistance to movement of earth masses: soil reinforced, soil filled, gabion retaining walls, pile rows, dowels
	Human activity	Special regime in the erosion-landslide zone: fencing of construction works, operation mode

2. Technical solutions of shell structures made of composite nanomaterials

The Department of PSGiF developed technical solutions using soil-reinforced and soil-filled shell structures made of composite nanomaterials, which can be created in cramped conditions by new methods of their construction and an application for inventions has been filed to eliminate such natural and technogenic phenomena (Figure 1).

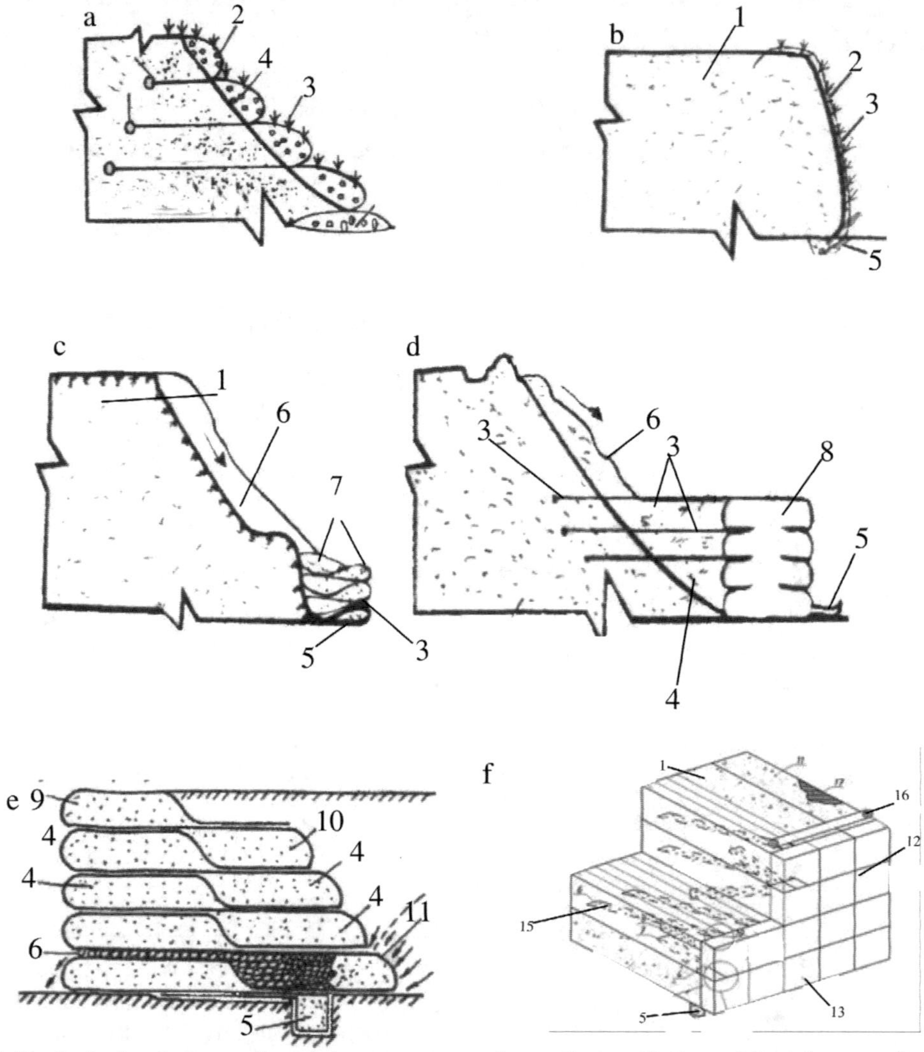

Figure 1. Technical solutions of retaining structures. 1 – soil massif; 2 – gabion element; 3 – arm tape; 4 – placeholder; 5 – drainage; 6 – moving soil massif; 7 – geotextile; 8 – fender; 9 – the front shell; 10 – internal soil-filled shell; 11 – perforation for removal of the filtration stream; 12 – front wall; 13 – front element; 14 – rectilinear arm tape; 15 – inclined arm tape; 16 – soil-filled shell

2.1. Calculation justifications

When designing retaining structures from soil-reinforced elements (Figure 1), the maximum horizontal force T, which keeps the slope in equilibrium for given soil characteristics (approximately for a triangular distribution of soil pressure), should be determined by the following relationship:

$$T = \frac{1}{2} \cdot K_a \cdot \gamma \cdot H^2,$$

(1)

where $K_a = tg^2\left(\frac{\pi}{4} - \frac{\varphi}{2}\right)$ is the coefficient of active soil pressure;

γ is the specific gravity of soil, kN/m3; H is the height of reinforced layer, m; φ is the angle of internal friction of the embankment.

When using straight and inclined arm tape, the calculation of the forces in them can be carried out according to the following relationship:

$$T_0 = T_1 + T_2 = \int_0^{l_1} 2 \cdot b_1 \cdot m_1 \cdot f_1 \cdot \gamma \cdot h \cdot dl_1 + \int_0^{l_2} 2 \cdot b_2 \cdot m_2 \cdot f_2 \cdot \gamma \cdot h \cdot dl_2 \qquad (2)$$

where T_0 is the total tensile force in all layers of reinforcement corresponding to the collapse surface, kN/m; T_1, T_2 is the tensile forces in rectilinear and obliquely located arm tape, kN/m; b_1, b_2 is the width of a straight and inclined arm tape, m; m_1, m_2 is the number of reinforcement strips per 1 m soil reinforced wall; H is the height of the considered layer of soil over armor tape; f_1, f_2 is the rebar coefficient of soil, where $f = tg(\varphi)$ (φ of the soil internal friction); l_1, l_2 is the straight and inclined arm tape length, m; γ is the bulk density of the embankment t/m3; k is the coefficient taking into account the slope of the arm tape.

Based on Experimental studies of soil-reinforced structures, taking into account patent No.2444589, were conducted under the supervision of the doctor of technical science Kasharina T.P. by graduate students Prikhodko A.P. and Kundupyan K.S. in a flat deformation tray made of organic glass with a working space of 0.8 * 0.1 * 0.6 m, while dry sand was used as a backfill and ground-reinforced foundation [5-11].

The sizes of the front wall (150, 100, 50 mm) and the front element (50, 30, 10 mm), the length of reinforcing tapes -l =0,7H, where H is the total height of the moving wall, were changed.

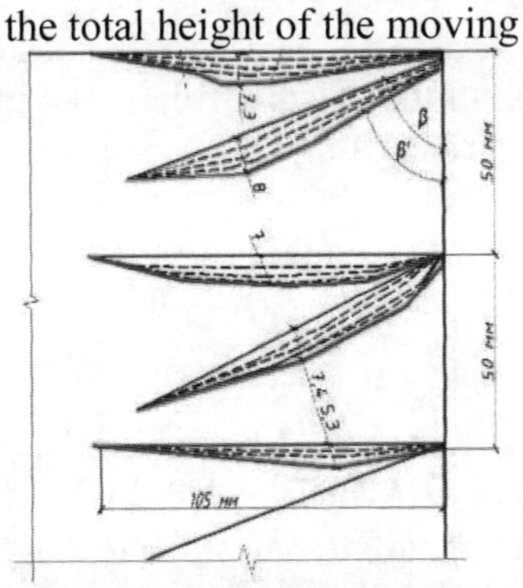

Figure 2. The diagram of fixed deflections of horizontal and inclined reinforcing tapes, obtained using the method of photometry

Figure 2 clearly shows that inclined reinforcing tapes experience the greatest deformation and they appear at 2/3 H. This allows the most efficient use and adjustment of the parameters of the ground-reinforced mass.

The ground-filled elements with sorbents, purifying groundwater and surface water, are used as drain anchors. They are calculated according to the empirical dependencies with a load of up to 120 kPa. The deformation of the ground-filled shell was $\varepsilon = 0{,}3$ mm, and its tension is determined by the dependencies [12-18]:

$$T = \frac{(h+y)(\{1-a\}+\{1-am\}y'^2)}{y''} + +(1-am)\left(hy+\frac{y^2}{2}\right), \quad (3)$$

where: T is the tension force in the shell,

$T = \frac{T_0}{\gamma_н b}$, kN/m.

The equation describing the shape of the shell is the elastic equation of the second kind:

$$y = \left(1-\sqrt{1-\frac{\sin^2\varphi}{k^2}}\right)h, \quad (4)$$

where: φ is the angle of internal friction of the backfill soil, degrees; k is the modulus of elliptic integrals.

The first and second derivatives of equation (3) take the following form:

$$y' = \frac{h\sin\varphi\cos\varphi}{k^2\sqrt{1-\frac{\sin^2\varphi}{k^2}}} \quad y'' = \frac{h(k^2\cos 2\varphi + \sin^4\varphi)}{k^4\sqrt{\left(1-\frac{\sin^2\varphi}{k^2}\right)^3}}. \quad (5)$$

On the basis of the obtained dependencies, a program was compiled for calculating a closed ground-filled shell under load [1].

3. Conclusion

The study results are currently used in the design of the retaining structures to protect the urban area. The above calculation justifications are necessary to create simulation models.

References

[1] Kasharina T P, Kasharin D V, Prikhodko A P and Zhmaylov O V 2012 Pat RU 2444589МПК E02B 3/06 *Ground-reinforced structure and method of its construction* declare 07/26/2010; publ. 10.03.2012 Bull. No.7

[2] Kasharina T P, Kasharin D V, Klimenko M Yu, Kundupyan K S and Sidenko E S 2016 Pat RU 2604933 E02B 3/02 (2006.01) E02B 5/00 E02B 8/06 B82Y 30/00 *Installation of the protective system of urban development and the method of its construction* declare 02.26.2015; publ. 09/20/2016 Bull. No. 35

[3] Kasharina T P 2000 *Perfection of structures, methods of scientific justification, design and technology of erection of lightweight hydraulic structure*. Abstract of Doctor's degree dissertation (Moscow) p 56

[4] Vidal H 1978 The development and future of reinforced earth. *Proceedings of Symposium on Earth Reinforcement, Geotechnical Enfineering Division of American Society of Civil Engineers* (Pittsburgh, USA) p 1

[5] Rudov-Clark S, Lomov S V and Bannister M K 2003 Geometric and mechanical modelling of 3D woven composites *Materials of the 14th International Conference on Composite Materials* (San Diego, USA)

[6] Juran I, Schlosser F, Long N and Legeay G 1978 Full scale experiment on reinforced earth abutment in Lille *Proceedings, Symposium on Earth Reinforcement, ASCE Annual Convention* (Pittsburg, USA) pp 586-584

[7] Matthews F L and Rawlings R D 1999 *Composite materials: engineering and science* (Oxford, Alden Press) p 470

[8] Kasharin D V 2018 Structures to protect the frame and soil reinforcement structures made of composite materials *Fundamental and applied issues of geotechnics: new materials, structures, technologies and calculations: a collection of annotations* (Saint-Petersburg: SPSUKU) p 60
[9] Ingold T S 1981 Soil reinforcing systems in the United Kingdom *Highways and Public Works* **49 (1858)** p 1620
[10] Kasharina T P, Zhmaylova O V and Glagolieva A S 2010 *Recommendations for use in low-rise construction of soil-filled elements with reinforcement of foundations on technogenic soils* (Rostov on Don: Yuzhvodoproekt) p 23
[11] Kasharina T P and Kasharin D V 2017 Application of shell structures made of composite nanomaterials *Bulletin of Perm National Research Polytechnic University. Construction and architecture* **8 (3)** pp 34-40
[12] Kasharina T at all 2012 *Creation of protective mono-envelope systems of artificial bases and foundations of buildings and structures and a device for its implementation* Application No.2012108682 of March 6, 2012, state registration number. Reg. 03/03/2014.
[13] Kasharina T P, Zhmailova O V and Glagoleva O S. 2009 *Ground-filled shell. Computer program* 2010610995 (No.2009616940)
[14] Khuberian K M 1956 *Rational forms of pipelines, reservoirs and pressure transitions* (Moscow: Gosstroyizdat) p 206
[15] Kasharina T P and Kasharin D V 2011 Use of soil-filled bases from composite materials for water-retaining structures. *Foundations of deep foundation and problems of the development of the underground space. Materials of the international conference* (Perm) p 395-401
[16] Kasharina T P, Zhmailova O V and Glagoleva A S 2009 Analysis of theoretical dependencies for calculating soil-filled shells *Science, Technology and Technology of the 21st Century* pp 346-350
[17] Floss R and Thamm B R 1976 Bewehrte Erde - Ein neues Bauverfahren im Erd - und Grundbau *Bautechnik* **53 (7)** pp 217-226
[18] Kasharina T P 2018 New technological solutions for structures using composite materials. *Fundamental and applied issues of geotechnics: new materials, structures, technologies and calculations: a collection of annotations* (Saint-Petersburg: SPSUKU) p 60

Quality ecological assessment of the urbanized territories' environment

L A Kaverzina[1], V A Nikiforova[1] and V V Dorofeeva[2]

[1]Bratsk State University, 665709, Makarenko str., 40, Bratsk, Russian Federation
[2]Kaliningrad state technical university, 236022, Ave. Soviet, 1, Kaliningrad, Russian Federation

E-mail: Dekanfps@mail.ru

Abstract. This article shows the results of a socially significant research, the relevance of which is constantly growing in current conditions, due to the impending global crisis. Research objectives are to develop theoretical knowledge in sphere of quality ecological assessment of the urbanized territories and identification of opportunities for their practical use in certain regions. Authors have systemized factors that affect on the quality of urban environment and identified main approaches, which are used in implementing of the quality ecological assessment of the urbanized territories' environment. Results of ecological and hygienic assessment of air quality of Irkutsk Region's (Russia) cities have been introduced as a part of the author's methodology. Main sources of pollution and their impact on atmospheric pollutant emissions have been identified. Urban areas with high level of air pollution have been revealed. Based on the findings, set of measures has been proposed to restore ecological balance in Irkutsk region, the implementation of which will improve the environmental conditions of atmospheric air as one of the criteria of quality assessment of urban territories. Methods of systemic and comparative analysis, expert opinions and social polls as well as ecological monitoring of the environment have been used during the research

1. Introduction

Quality of urban environment is an object of close attention from scientists, society,people and authorities at various levels. This problem is especially relevant for urbanized territories requiring sustainable development and increase of their competitiveness [1, 2]. A huge number of modern economists and practitioners study the issues of forming comfortable urban environment, its quality assessment and process of urban environment management [3-7].

Economical, biological and ecological factors have significant influence on the urban environment quality. The ecological principles of the Russian Federation policy developed for the next decade (until 2030) determine preservation of favorable environment as one of the strategic goals. Consequently, the minimization of negative influence on the environment with various factors of urbanized territories is the priority task [8].

Population of urbanized territories is extremely affected by techno genic influence. This fact allows to conclude that the reduction of negative impact for these territories is very relevant. Moreover, relevance of the quality assessment of urban environment is increasing too. It is especially important for disadvantaged regions, which include Nizhny Novgorod region, Orenburg region, Irkutsk Region,

Chelyabinsk region and Sverdlovsk region (according to the all-Russian public organization "Green Patrol" in 2019).

In current research, we will take a closer look at the situation in Irkutsk region where the largest power engineering, non-ferrous metallurgy and timber enterprises in Russia are situated. At the present time, the priority objective for Irkutsk region cities is to create favorable environment and support ecological balance of the territory. It confirms by results of the research, which are shown at [9, 10].

Development of cities in terms of forming the comfortable urban environment which includes social and economic, financial, industrial, personnel and other aspects, determines level of development of regional economy as a whole. Choosing a place to live, person takes into account different conditions – climatic, educational, social etc. But ecological conditions are the most important [11, 12]. In this context, we can note that more than 70% of the Russian Federation population is concentrated in cities, development of which leads to deterioration of the urban environment.

In modern conditions, there are serious ecological disruptions at the urban territories of Russia. These problems are primarily associated with environmental pollution [8, 13, 14], which negatively affects urban people's health [15-17].

An important role in improving the quality characteristics of the environment in urban territories belongs to the development of social and ecological oriented urban-planning [18, 19].

Purpose of the research is to assess the level of chemical air pollution of the urbanized territories of Irkutsk region as a part of the urban environment quality.

2. Research methods

During the current study, various research methods have been used: statistical methods; method of expert assessment and sociological questioning; general scientific methods; system analysis method; experimental methods, which allow to make complex approach to investigate the article's problem. Objects of this research are individual municipalities of Irkutsk region. Quality of the urban environment has been assessed according to the information from monitoring sources, which are situated in municipalities.

For achieving the target of the current research and solving assigned tasks, the necessary statistical information and laboratory data has been collected, including those presented in [20], for the period from 2005 to 2018. A special attention has been paid to the analysis of atmosphere pollution, because this aspect has the greatest influence on forming of comfortable urban environment.

In this way, the ecological monitoring of environment has been used as a main method of the research; analysis and data processing, comparative and systematization methods has been used as additional.

3. Results

Assessment of the quality of urban environment as a part of the current research has been implementing in the territory of the Irkutsk region characterized by special orographic and climatic conditions.

Complex ecological and hygienic researches suggest dynamics analysis of air pollutant emissions. The research period is 14 years (2005-2018). The monitoring of atmospheric air has been implementing in 9 municipalities (Angarsk, Bratsk, Irkutsk, Shelekhov, Ust-Ilim, Usolye-Sibirskoe, Cheremkhovo, Zima) (Figure 1). It is important to notice that every year the volume of pollutant emissions in cities of Irkutsk region from stationary sources has been carried out at the level from 502,3 to 720,3 thousand tons. The largest volume of the pollutant emissions is specific for Angarsk (from 130 to 265,6 thousand tons a year), Bratsk (from 92,9 to 124,3 thousand tons a year) and Irkutsk (from 58,5 to 70,6 thousand tons a year). The current situation in terms of pollutant emissions has led to the fact that Siberian Federal District ranks third after Krasnoyarsk territory and Kemerovo region.

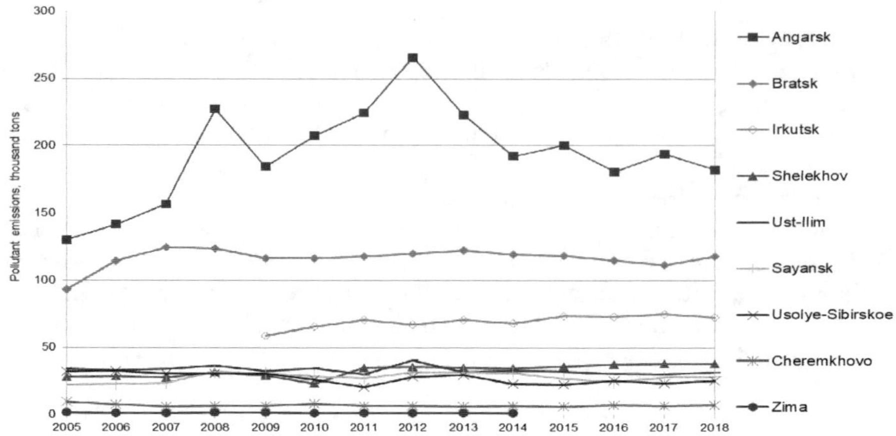

Figure 1. Air pollutant emissions into the atmosphere of Irkutsk region.

The range of characteristics has assessed quality of the atmospheric air in region's cities: data about the volume of impurity emissions; chemical pollutant conversions; meteorological parameters, taking into account impurity transfer and dispersion; integrated index of air pollution, etc. Table 1, compiled according to [20], shows the information for 2018 about the state of atmospheric air in Irkutsk Region.

There are many industrial enterprises in the territory of Irkutsk region. Their activity, taking into account the implementing of technological processes, means pollution of environment. In result, many cities are in the category of unfavorable territories.

At the present time, the process of construction of industrial facilities continues. In particular the construction of Taishet aluminum smelter which will increase the environmental burden in the region. To support ecological and economic balance of the region in terms of solving the problems cost-benefit analysis, it can be used approaches and indicators proposed in [21, 22].

Table 1 shows that the most cities have the level of air pollution rated as "very high». In almost all municipalities, the average annual concentration of individual impurities exceeds the maximum permissible concentration: in Bratsk the excess is noted for 4 types of impurities, in Irkutsk, Angarsk and Shelekhov for 3 types, in Baikalsk for 2 types, in the rest for 1 type.

Table 1. The state of atmospheric air in Irkutsk region in 2018.

City	Indicators	
	Air pollution level	Exceeding annual average maximum permissible concentrations (MPC)
Angarsk	«very high»	nitrogen dioxide -1,6; benzpyrene - 5,8; ozone - 1,1.
Bratsk	«very high»	carbon disulfide - 1,8; suspended substances - 1,5; benzpyrene -6,2; formaldehyde - 2,0;
Irkutsk	«very high»	benzpyrene - 7,8; suspended particles PM 10 –1,3; suspended substances - 1,5.
Zima	«very high»	benzpyrene -14,6
Svirsk	«very high»	benzpyrene -14,5
Usolye-Sibirskoe	«very high»	benzpyrene - 10,1
Cheremkhovo	«very high»	benzpyrene - 8,7
Shelekhov	«very high»	ozone -1,3; hydrogen fluoride - 1,4; suspended particles PM 10 - 1,3
Baikalsk	«high»	ozone - 2,0; benzpyrene -1,7
Sayansk	«high»	benzpyrene -3,2

Exceeding concentration of carcinogenic substances into the atmospheric air (formaldehyde and benzpyrene) is observed in 9 cities out of 10 presented in Table 1.

During a general assessment of the urban environment state, the impact of motor transport in pollutant emissions in Irkutsk region has been revealed. It causes a high level of pollution index (table 2). Despite the growth in recent years, the number of motor transport in Irkutsk Region, the pollutant emissions remain almost at the same level of 377,8 thousand tons. However, there is a growth in the city of Bratsk in 2018 of 80% compared to 2005. In Sayansk, Angarsk, Ust-Ilim, Cheremkhovo the level of pollution almost do not change.

The analysis conducted by the authors allows concluding that such ecological indicator of the urban environment as air quality, in large Irkutsk Region municipalities for many years remains unsatisfactory. The consequences of such negative trends lead to poor health and the formation of environmentally related pathologies. Hence, there is an objective necessity of development of the methodology of the quality ecological assessment of the urbanized territories' environment that allows to consider individual characteristics of assessing territories, make conclusions about the state of area and develop activities for solvation of the revealed problems in researched territories.

Table 2. Pollutant emissions into the atmosphere from motor transport.

City	Pollutant emissions, thousand tons		
	2005	2010	2018
Angarsk	32.08	32.53	32.20
Sayansk	5.10	5.79	5.80
Bratsk	21.88	39.54	39.80
Irkutsk	-	92.80	65.10
Ust-Ilim	13.93	13.62	13.30
Usolye-Sibirskoe	10.09	9.65	16.80
Cheremkhovo	12.67	12.78	12.10
Shelekhov	9.86	9.72	10.20
Irkutsk region	377.88	373.87	377.80

The methodology proposed by the authors includes the following stages:

– detection of factors that influence the quality of urban environment and removal of insignificant factors;

– analysis of importance of the remaining factors in terms of main goals and objectives for development or existing territory problems;

– detection of necessity for conducting an ecological assessment of the quality of urban environment;

– collection and processing of information about ecological state of investigated urban territory, clarification of the most important ecological problems and detection of the most significant factors and sectors of economics that affect the ecological part of the investigated territory;

– situation forecast in area of the changes of urban environment quality;

– development of activities for improvement the quality of investigated urban environment.

This way, the environment of urbanized territories constantly suffers from negative influence of industrial economics sector. In hard modern conditions state, regional and municipal authorities must make a complex approach for solving of the issue of forming a favorable urban environment in the territories entrusted to them. Otherwise, it will negatively affect natural and anthropogenic sustainability and may cause negative social and economic consequences at the micro, meso and macro levels.

4. Discussion

Issues of the quality ecological assessment of the urbanized territories' environment are debatable and require practical confirmation. Close attention is paid to discussing the problems about influence of city noise, pollutant emissions of industrial enterprises and water quality on people's health [11,14]. Different ecological and epidemiological researches are implemented [23]. Special attention is paid to assessment of the risks related to influence of environmental factors on health of people who live in urbanized territories [13, 16, 24, 25]. This problem is the most relevant for children [12, 15].

In hard modern conditions, it is necessary to reduce the negative impact of city environment factors on health of urbanized territory population while increasing the objectivity of quality ecological assessment of urban environment.

Conclusion

The research conducted by the authors has allowed to solve the following tasks in sphere of quality ecological assessment of the urbanized territories' environment:

1. Possibility of an integrated assessment of the quality of the urban environment of cities with overloaded industrial infrastructure has been considered. The role and significance of ecological assessment of specified territories have been determined.
2. Main factors that determine the quality of industrial city environment have been revealed. These factors have been systematized and characterized.
3. Formal ecological assessment approaches of urbanized territories' environment have been justified
4. The methodology of quality ecological assessment of urban environment that allows to take into account individual features of estimated territories (natural and climatic, demographic, anthropogenic, technogenic, ecological, social and economic features) has been developed.
5. Comparative analysis of ecological characteristics for individual municipalities of Irkutsk region to assess the quality of these territories environment, reveal existing problems and develop activities for their liquidation, has been conducted.

Practical value of the scientific research conducted by the authors is to develop the system of administrative, technological, planning and sanitary measures aimed to solving revealed problems of Irkutsk region in field of formation quality urban environment. Results of the research and developed proposals may be used by regional and municipal authorities for restoration of ecological balance in territory of Irkutsk region.

References

[1] Musikhina E A and Khokhrin E V 2013 *Bulletin of Irkutsk state technical University* **4 (75)** pp 97-101
[2] Vyatkina B M 2012 *Bulletin of Irkutsk state technical University* **4 (65)** pp 74-79
[3] Mitichkina A E 2014 Bulletin of Irkutsk state technical University **8 (91)** pp 94-100
[4] Yaskova N Y 2013 *Bulletin of Irkutsk state technical University* **10 (81)** pp 380-382
[5] Shtepa E A 2014 **6 (89)** pp 134-137
[6] Paraschuk E M and Rubanov V G 2014 *Bulletin of Irkutsk state technical University* **3 (86)** pp 12-20
[7] Moiseev Yu M, Kartashova K K and Melodinsky D L 2019 *News of higher educational institutions. Construction* **8 (728)** pp 104-115
[8] Klepikov O V, Samoylov A S, Ushakov I B, Popov V I and Kurolap S A 2018 *Hygiene and sanitation* 97 **8** pp 686-692
[9] Dmitrieva Yu N 2016 Izvestiya Irkutsk state University.Series: Earth Sciences **17** pp 34-45
[10] Aleshina I N and Turkina N G 2012 Izvestiya Irkutsk state University. Earth Science series 5 **2** pp 265-276
[11] Rakhmanin Yu A and Levanchuk A V 2016 *Hygiene and sanitation* 95 **11** pp 1021-24
[12] Chanchayeva E A, Gvozdareva O V and Gvozdarev A Yu 2019 *Human ecology* **12** pp 12-19

[13] Vekovshinina S A, Klein S V, Zhdanova-Zaplesvichko I G and Chetverkina K V 2018 *Hygiene and sanitation* 97 **1** pp 16-20
[14] Kopytenkova O I, Levanchuk A V and Eremin G B 2019 *Hygiene and sanitation* 98 **6** pp 613-618
[15] Nikiforova V A 2016 *Hygiene and sanitation* 95 **9** pp 841-847
[16] Karelin A O, Lomtev A Yu, Volkodaeva M V and Eremin G B 2019 *Hyena and sanitation* 98 **1** pp 82-86
[17] Dolgushina N A and Kuvshinova N A 2019 *Human ecology* **6** pp17-22
[18] Buzyrev V, Nuzhina I and Zolotareva M *Social-ecological priorities of town-planning activities on the urbanized territories* MATEC Web Conf. (ESCI 2018) 193 (Ho Chi Minh City 02-05 March 2018) p 01007
[19] Nuzhina I P, Zolotareva M V and Vasil'eva J V 2018 *Integration of urban developers with regard to social and environmental responsibility* MATEC Web Conf. (YSSIP-2017) 143 (Tomsk Russian Federation 25-27 October 2017) p 04010
[20] State report 2019 "On the state and environmental protection of the Irkutsk region in 2018" (Irkutsk: Megaprint LLC)
[21] Kogan A 2015 *Journal of Advanced Research in Law and Economics* VI **1(11)** pp 112-120
[22] Kogan A 2014 *Journal of Advanced Research in Law and Economics* V **1(9)** pp 15-22
[23] Petrov S B, Onuchina E N and Petrov B A 2012 *Human ecology* **3** pp 11-15
[24] Ushakov I B, Klepikov O V, Popov V I and Samodurova N Yu 2017 *Hygiene and sanitation* 96 **9** pp 904-908
[25] Rosenthal O M and Alexandrovskaya L N 2019 *Hygiene and sanitation* 98 **5** pp 563-569

IOP Conf. Series: Materials Science and Engineering **911** (2020) 012017 doi:10.1088/1757-899X/911/1/012017

Formation of a comfortable urban environment, taking into account the reconstruction, restoration of buildings and structures (on the example of the Kaliningrad region)

V I Pustovgarov[1]

[1]Immanuel Kant Baltic Federal University, 14 Nevskiy St, Kaliningrad, 236016, Russian Federation

E-mail: viktorpustovgarov@mail.ru

Abstract. The issues of reconstruction of the historical environment of cities are one of the most acute when considering the problems of development of urbanized territories. A significant number of examples of the loss of historical monuments make it especially careful to the preserved architectural heritage. The process of revitalizing the planning of the territory in the area of the South (Yuzhniy) Station in the city of Kaliningrad is due to the incompleteness of the town-planning ensemble, which began to form at the end of the 19th century. Urban planning reconstruction of the territory is a complex of necessary scientific research and the preparation of a historical and cultural reference plan, which is defined by the Federal Law "On cultural heritage sites (historical and cultural monuments) of the peoples of the Russian Federation" [1]. As a result of studying the stages of urban development of the historical territory of the city and analyzing its architectural and spatial organization, conceptual proposals for its development were developed.

1. Introduction

The area adjacent to the Yuzhniy railway station in Kaliningrad is one of the most iconic urban areas. This territory currently houses the Main Railway Station, serving both suburban and long-distance trains, a bus station, shopping centers, residential and public buildings, and infrastructure facilities. This territory, to a certain extent, personifies the image of the city, in the form of a unique example of cultural, historical and urban planning heritage. In the process of research to identify the historical and cultural characteristics and value of elements of the urban environment, the main problems of the study area were identified:

- randomness of the territory development;
- incompleteness of the ensemble of the station square;
- conducting construction in the absence of scientific research in the field of methods of regenerating the historical environment and restoration of individual historical buildings;
- construction in historic green spaces.

In addition, adjacent streets are problematic, as their traffic capacity is insufficient.

In the General Plan of the city of Kaliningrad (2006), developed in accordance with the Urban Planning Code of the Russian Federation [2], it was determined that the territory of the Yuzhniy Station should be consolidated as an urban planning element of the cultural heritage. Also, objects of cultural heritage must be endowed with rather strict regulations for use: from the requirement for the restoration of facades to the complete restoration of buildings, including interiors and historical

IOP Conf. Series: Materials Science and Engineering **911** (2020) 012017 doi:10.1088/1757-899X/911/1/012017

improvement. Contrary to these decisions, shopping centers were built on the territory not intended for development, which led to a significant deterioration in the perception of the entire ensemble of the Yuzhniy Station.

Architecture knows many phenomena when an idea is embodied, and moreover, only one for each case. This is the regularity. Various external interpretations are possible, but there is only one embodiment of a given idea in the material essence of this idea. The power of architectural excellence is that this external interpretation, i.e. transmission of expression, thought, accurately set out the tectonic content of the idea. Let's not forget that architecture is a presentation of tectonics [3].

2. Disclosure and preservation of the social value of the historical heritage and its inclusion in the modern life of the city

In the project of zones for the protection of historical and cultural monuments of the city of Kaliningrad, carried out by the Scientific and Design Institute for Spatial Planning "ENKO" (St. Petersburg) in 2002, the main stages of the city's urban development were determined:

- Order period – middle of 13th century - the beginning of the 16th century. Construction of the Castle and 3 cities: Altstadt (1286), Lebenicht (1300), Kneiphof (1327);
- The period of Absolutism – early 16th century - the end of the 18th century;
- Imperial period – 19th century;
- Republican period – early 20th century;
- The period of urban planning stagnation after the Second World War – 1945 - mid-1960s;
- Soviet period – mid 1960-1991;
- Russian period – from 1991.

On early plans and maps up to the beginning of the 18th century, the study area is not included in the city boundaries - thus, we can conclude that there was no active development of it. The main thing that was characteristic of this place was that the road to the castle of the Teutonic Order of Brandenburg passed through it in the south-west direction, the ruins of which are still in the village of Ushakovo on the shore of the Kaliningrad Bay. In 1626-1634, work was underway in the city to build the first rampart defensive ring around the three cities of Altstadt, Kneiphof and Lebenicht. On June 13, 1724, the Prussian king Friedrich Wilhelm I signed a decree on the unification of three cities and suburban communities into a single city of Konigsberg [4]. The rampart fortifications included 11 bastions, 2 ravelins and provided for the entrance to the city through 9 gates. It is from the appearance of the fortress wall on the plans and maps that it becomes possible to uniquely identify the territory of the modern complex of the Yuzhniy Station. In the middle of the 17th century, here, at the intersection of modern Bagration and Zheleznodorozhnaya streets, there were: the Brandenburg bastion, the Haberberg ravelin - east of Leninsky Avenue, in the area of existing residential buildings.

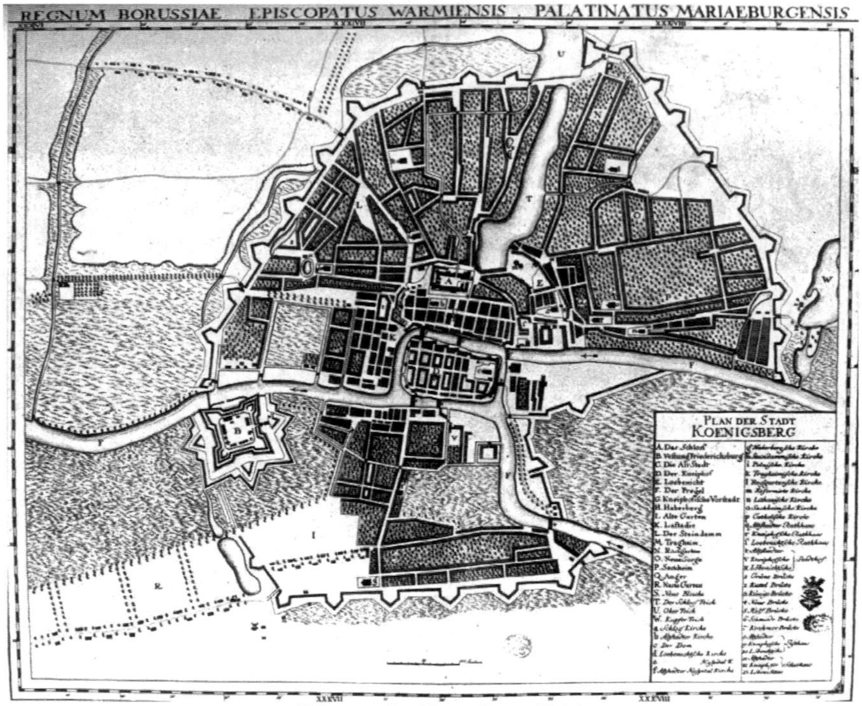

Figure 1. City plan, 1701.

Between the bastion and the ravelin, on the territory of present-day Kalinin Square, there was a cavalier, which was an "intermediate" fortification of the First Rampart Defensive Ring. Behind the fortress walls, that is, on the site of the modern track facilities of the South Station, there were arable fields and a mill. Inside the city walls, in the area of the modern House of Arts, the Haberberg Church existed since 1652. It is interesting to note that in 1783, on the advice of Immanuel Kant, who was baptized in the Haberberg Church, the first lightning rod in Konigsberg was installed in the building of the church.

In 1840, the Konigsberg Landtag made a petition to the king to build a strong fortification around Konigsberg. Construction work began on October 15, 1843 with the ceremonial laying of the Kronprinz defensive barracks [5]. During this period, the construction of the second defensive ring with new powerful engineering fortifications (forts, bastions) was carried out. The basis of the urban planning culture of this period lies in the appearance of a two-part composition of the spatial planning structure of the city, which already consists of two types of urban planning grid:

Type I - urban planning discipline "hippodamov" plan of the central part of the city;

Type II - the radial structure of the inner city, already formed on the basis of historical, natural and trade roads leading to the castle.

At the beginning of the 20th century, the expansion of the city's territories and the growth of industry continued. "Until 1930, a new Northern (Severniy) railway station was built for the Kranz, Labiaus and Samland railways and the new Main railway station in the south of the city, with a large station square stretching all the way to the Haberberg church. In contrast to the old East and South railway stations, the new Main railway station was a gateway station. From here, trains departed to Labiau and Pillau, the first via Gare du Nord, and the second via Hollender Baum (Dutch Gate). Thus, the traffic flow went from the Kaiserstrasse street to the Vorstadt Langgasse. A technical novelty was the two-story railway bridge over the Pregel, built in 1926" [6].

As Professor Dr. Baldur Koster writes: "The planning of railway structures began as early as 1896. In this sense, the demolition of the fortifications that took place had a positive effect: the demolition of existing buildings was not required for the new main station and the tracks of the freight station. Quite the contrary, for the former ramparts in the south-east between the Haberberg bastion and the

Friedrichsburg fortress, they could find another use. Railway station square (Reichsplatz) - extremely large by the standards of Konigsberg - from the then positions was laid in the most optimal way for traffic. In front of the long facades of the railway station building and the adjacent administrative building (there is a second exit here) a long square appeared, where only trams stopped" [7]. The station building was designed by the high-rise building department of the Berlin Railways Administration with the participation of the architect Cornelius in 1929. At that time it was one of the largest railway stations in Europe.

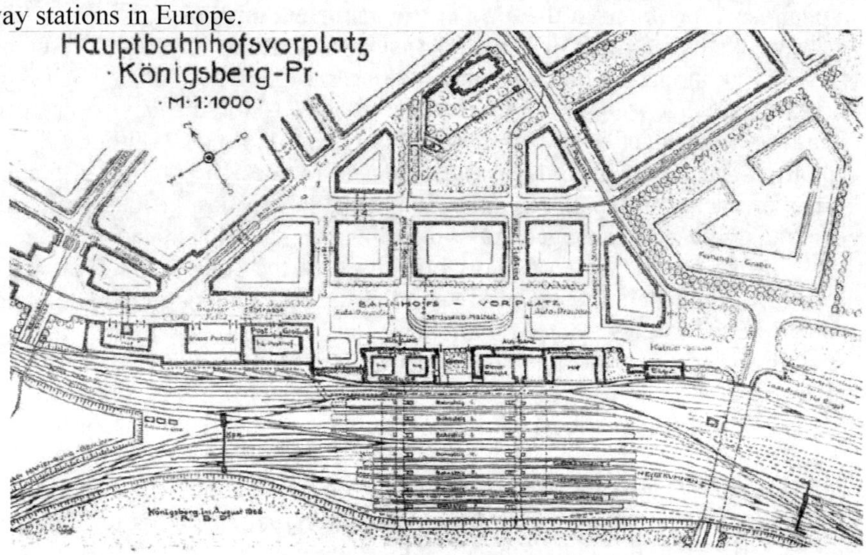

Lageplan Hauptbahnhof Königsberg (Pr.). M. 1:4000.

Figure 2. Development plan for the main station complex and the station square. 1929-1940 [11]

Figure 3. The complex of buildings of the Main Station 1929-1940 [11]

In the first post-war period, the element of continuity of the cities of Konigsberg - Kaliningrad was still preserved, but already the following general plans of the city, developed in the 60s, provided for a complete rejection of the city's planning structure that had been historically established and established over the centuries. As a result, an ideological orientation was adopted to completely ignore the previous architectural and urban planning component. This led first to the appearance in one place of two completely different cities, then to the collision of their architectural and urban planning cultures, and later to a complete change in the structure, character, scale and image of the city. The key concept of this time is rationalism, as a result of which the utilitarian becomes an aesthetic category. An ideally rational building within an ideally rational city in the urban planning ideology is a microdistrict. This is the basis of the urban planning culture of the seventh historical stage of the city's development.

3. Conceptual proposals for the revitalization of the historical territory

When solving the problems of reconstruction of historical districts in more preserved and more valuable areas, it is appropriate to maintain the existing morphotypes, preserve the laws of their relationship. In other cases, it is possible to consider proposals for the development of emerging features, and even for the introduction of new architectural themes into the building [8].

A certain remoteness in time with today's assessment of the design solution of the station square allows it to be perceived in an expanded cultural context, where design ideas often could not be realized, but were transformed and changed under the influence of various factors. At the same time, it is proposed to restore the design scale of the public space of the square, create a single ensemble, which will be formed by the reconstructed buildings of shopping centers and new buildings of the bus station, hotel, public and business centers, buildings of the temple complex and the complex of buildings of the railway station. The planning organization of the territory according to the plan obeyed strict geometry and stable balance with the self-sufficiency of each node of the composition. Community is also achieved by some kind of unity of action, where public transport is one of the main participants.

As a result of the studies carried out, taking into account also the conclusions from the master plans of Kaliningrad in 2006, 2016, it is possible to formulate conceptual proposals for the revitalization of the studied historical territory in the area of the Yuzhniy station:

1. The identified development plan of the early 20th century is the most durable and basic element of the spatial structure.
2. The second element of the reconstruction is the idea that the scale of development is the lasting value of this territory.
3. Reconstruction of the architectural-spatial, historical-cultural and landscape originality of the territory.
4. When carrying out the reconstruction of this territory, it is necessary to form a complete compositional urban planning ensemble, the main objects of which should be:

- reconstructed residential development and development of multifunctional buildings in the area adjacent to Kalinin Square;
- development of a transport hub, taking into account the optimization of the use of the territory and the road network;
- separation of traffic flows to the Yuzhniy Station, to the bus station, transit transport, public intracity transport, intercity and international transport;
- construction of underground parking lots, underground and overground pedestrian crossings;
- expansion of the carriageway of streets;
- expansion of the territory of the bus station.

5. It is proposed to clarify the zones of protection of objects of cultural heritage of regional significance and their territory in accordance with the Regulation on zones of protection of objects of cultural heritage (historical and cultural monuments) of the peoples of the Russian Federation [9].

4. Conclusion

The tendency in the Kaliningrad urban studies of recent years is associated with a revival of interest in the main central public spaces, a rethinking of their role and significance in the structure of urban space and the development of models of their evolution in new economic conditions. Improvement has become synonymous with reconstruction. Everyone is well aware of the severity of transport, environmental and resource problems. Involvement in commercial turnover and, consequently, in the process of revitalization of more and more areas of the historical urban environment, is one of the important tasks facing the authorities and the urban community. The considered territory of the Yuzhniy Station is one of the centers of attraction for people. Along with solving the problems of the transport hub, the environmental approach should prevail as the main priorities. In this situation, high-quality improvement is designed not only to provide psychological comfort, but also to preserve, emphasize and reveal the symbolism of the place, its inner atmosphere and individuality.

The attitude to the past becomes creative only when the architect is able to understand its inner meaning [10]. The square near the Yuzhniy Station has been preserved as a vast open space, although it was somewhat changed after the last construction of a number of buildings, but its public, socio-cultural essence as a square has practically disappeared. This implies one of the main conclusions that the square should revive its role as an attractive place for townspeople and for tourists, become a powerful utilitarian frame of the symbolic dominant of the Yuzhniy Station building in accordance with the identified planning decision of the 20s years of the last century.

References

[1] Federal Law of 25.06.2002 N 73-FZ 2005 *On cultural heritage sites (historical and cultural monuments) of the peoples of the Russian Federation* (Moscow)
[2] *Russian Federation. Laws. Town-planning code of the Russian Federation* Electronic resource: federal law. Reference and legal system "Consultant Plus"
[3] Pasternak K V 2007 *Civic centers of the cities of the ancient world of the Middle Ages and classicism* (Moscow: Stroitelstvo i biznes Press) p 389
[4] Albinus R 1988 Lexicon of the city of Konigsberg and its surroundings *Leer*
[5] Gubin A B and Strokin V N 1991 *Ocherki istorii Konigsberga* (Kaliningrad: Калининград: Kaliningrad book publishing house) p 190
[6] Gause F 1994 *Koenigsberg in Prussia. The history of a European city* (Bitter) p 19
[7] Koster B 2014 *Koenigsberg. Today's Kaliningrad. The architecture of the German time* (Kaliningrad: LLC Zhivem) p 272
[8] Shchenkov A S 2013 *Reconstruction of historic cities* (Moscow: Monuments of historical thought) p 420
[9] *Regulations on protection zones for cultural heritage objects (historical and cultural monuments) of the peoples of the Russian Federation* 2008 (Moscow: "RG" - Federal issue No. 4653)
[10] Gidion Z 1984 *Space, time, architecture* (Moscow) p 17
[11] *Visual archive of East Prussia* Electronic resource: https://www.bildarchiv-ostpreussen.de

IOP Conf. Series: Materials Science and Engineering **911** (2020) 012018 doi:10.1088/1757-899X/911/1/012018

Risk assessment of investment projects implemented on the basis of project financing

I V Bryantseva[1] and N V Voronina[2]

[1] Doctor of Sciences in Economics, Professor, Pacific National University, 136 Tihookeanskaya Street, Khabarovsk, 680035, Russian Federation

[2] Candidate of Sciences in Economics, Associate Professor, Pacific National University, 136 Tihookeanskaya Street, Khabarovsk, 680035, Russian Federation

E-mail: nat_vnv@mail.ru

Abstract. Risk assessment of investment projects still remains a difficult problem in the real investment practice. The development of project financing techniques and methods complicated this task, because new subjects of investment activity, participating in the implementation of the project and requiring consideration of their interests appeared. This, in turn, is seen in the redistribution of common project risks. In addition, new forms of financing create new risks. In this article the authors systematized the risks of project financing which exist for every participant of the investment process throughout the project lifecycle. This approach has been tested on a real investment project. Taking risk into account and risk assessment should allow all interested parties to make an informed decision about the advisability of participating in the project.

1. Introduction

Recently, due to the inappropriateness or impossibility of applying traditional project investment methods, there has been a need for new tools and forms of attracting investments. One of such forms is project financing, which differs significantly from traditional tools and makes it possible to put into practice on favorable credit conditions large-scale projects that require high costs.

This method of financing represents financing of investment projects, in which the source of repayment of investments is the income generated by the project in the future.

2. Materials and methods

A significant contribution to the study of the mechanism of project financing and the evolution of its forms was made by Peter K. Nevitt, Philip R. Wood, Frank J. Fabozzi, Richard R. Brealley, Stewart C. Myers, who primarily noted its isolation from the main activities of the company and the ability to minimize its risks [1, 2, 3, 4].

In their researches Shenaev V. N. and Irniyazov B. S. focus attention on the problems of risk sharing between participants of project financing [5]. However, despite the variety of studies in this area, modern risk management techniques for project financing have not yet been sufficiently formed. In our research we used abstract-logical, statistical-economic, economic-mathematical and expert research methods of risk systematization and assessment.

3. Results

The authors conducted a study that allowed them to systematize and provide an objective evaluation of the investment projects risks which are implemented on the basis of project financing. By conducting an expert survey among entrepreneurs, investors, scientists and specialists who are directly involved in the development and implementation of investment projects, as well as in consulting, we identified the types of risks and the degree of their influence at various stages of the implementation of investment projects. Distribution of answers to the question "What types of risks have a significant impact on investment projects implemented on the basis of project financing?" is given in Figure 1.

Among the risks that have a significant impact on investment projects which are implemented on the basis of project financing, most of the respondents indicated financial and economic risks. It is these risks that accompany any large project and affect the cash flows generated by the project, as well as its financial stability and solvency. In addition, according to respondents, legal and market risks also have a special impact on investment projects. Some respondents noted that some projects may have specific risks that are not included in the response options we offered them to choose from. This is due to industry-specific features of ongoing projects, as well as to the lack of an open mechanism for conveying the experience of using this method in investment projects in the constituent entities of the Russian Federation.

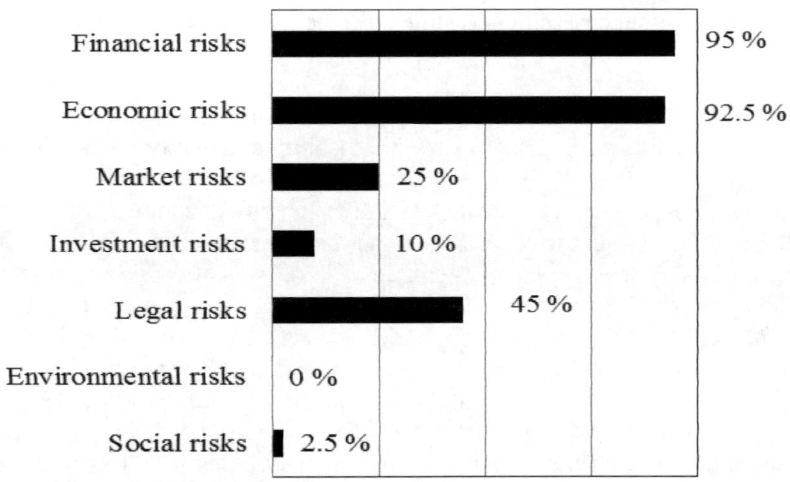

Figure 1. Distribution of answers to the question "What types of risks have a significant impact on investment projects implemented on the basis of project financing?"

For effective risk management, it is necessary to assess risks, so one of the questions was devoted to identifying methods that will help to objectively assess their size. Distribution of answers to the question "With the help of what methods is it necessary to assess the risks of investment projects implemented on the basis of project financing?" is presented in Figure 2.

Most of the respondents believe that it can be done more accurately by analyzing project development scenarios and by sensitivity analysis, however, to conduct a more objective risk assessment, it is necessary to use a combination of the presented methods. Also, about 20% of respondents expressed a special opinion that it is necessary to develop new techniques for assessing risks of investment projects implemented on the principles of project financing and to evaluate the specific risks of the borrower and lender separately. In domestic and foreign practice, there is such experience, for example, to use methods of mathematical modeling to assess the risks of projects financing [6].

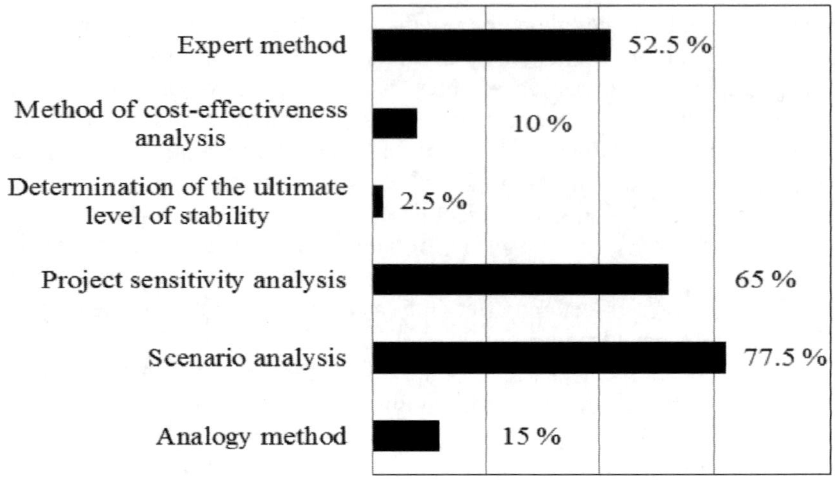

Figure 2. Distribution of answers to the question "With the help of what methods is it necessary to assess the risks of investment projects implemented on the basis of project financing?"

In special scientific literature, there is virtually no uniform classification of the risks of investment projects. Traditionally, the risks associated with the implementation of an investment project are divided into systematic (non-diversifiable) and unsystematic (diversifiable). Systematic risks include environmental and project risks (political, legislative, macroeconomic). Unsystematic risks include risks associated with the project itself and with the mechanism of its financing. Methods of analysis and development of measures to reduce systematic and unsystematic risks differ significantly [3, 4].

Some authors consider project risks as a whole, by which they mean the deterioration of the final indicators of the project efficiency arising under the influence of potential threats, uncertainty or incorrect project structuring [7, 8]. Project risks are determined by the fact that future project income is not a deterministic value that is known at the time of the decision to invest, as well as the amount of potential losses or lost revenues is unknown. Therefore, a thorough risk analysis is necessary in order to determine the risk profile of the investment project, ensure qualified risk management and develop effective techniques to reduce them at various stages of the project life cycle [9, 10].

Along with common risks that are typical for all investment projects, each project has its own characteristics, a specific contractual and financial structure and conditions for its implementation, and it is associated with industrial and regional risks specific to this particular project [11, 12, 13].

The study which we conducted allowed us to systematize and identify the risks of project financing, which are presented in Table 1.

Table 1. Systematization of risks of project financing.

Subject of the project	Identification of the risk connected with the activities of the subject
Customer	Funding risk is associated with the inability to attract the necessary funding within an acceptable time frame and on acceptable terms
Sponsor, Investor	Risk is associated with the lack of the necessary amount of free cash from sponsors and investors
General contractor,	Risk is associated with violations in the organization

Subcontractor	of construction and a delay in the facility commissioning
Project management company (project operator)	Operational risk is associated with the effectiveness of the organization in operation of the project site, its technical safety
Lender	Interest rate risk is associated with changes in interest rates
Special purpose vehicle	Risk of late payment of debt and a high level of leverage in the capital structure
Consumer	Risk is caused by market and qualitative changes in demand for project products
Competitors	Competitive risk is associated with the emergence of new construction organizations in the market
State authorities	Risk is associated with the creation of restrictive barriers by the state

Let us consider the process of risks systematization, identification and assessment using the example of a project for the construction of a block of flats that has 93 apartments in the city of Khabarovsk (Russia). The financial model incorporates a project financing scheme with full recourse on the borrower. The share of borrowed capital is 80% of all project capital costs.

In this project, seven types of risk were identified: the risk of sponsors and investors; general contractor risk; operational risk; competitive risk; funding risk; interest rate risk; ultimate customer risk.

The results of these risks assessment are given in Table 2.

Table 2. Risk event probabilities.

Name of risk	Risk probability		
	Low	Medium	High
Risk of sponsors and investors	0.15	0.25	0.35
Risk of the general contractor	0.20	0.35	0.50
Operational risk	0.05	0.10	0.15
Competitive risk	0.10	0.15	0.20
Funding risk	0.30	0.35	0.40
Interest rate risk	0.15	0.20	0.25
Ultimate customer risk	0.20	0.25	0.30

The conducted project risk analysis made it possible to make certain adjustments to the initial financial model of the project and on its basis to calculate the performance indicators, which are presented in Table 3.

When we took into account all the risks of the project financing which we researched we could identify weaknesses which the project owner had. If these risks arise, his participation in the project is ineffective. The weak position of one of the participants in project financing threatens the implementation of the entire project as a whole.

Table 3. Performance indicators before and after the adjustment of the financial model of the project.

Name of indicator	Before adjusting the financial model of the project			After adjusting the financial model of the project, taking into account identified risks		
	For the creditor (bank)	For the project	For the owner of capital	For the creditor (bank)	For the project	For the owner of capital
Net present value (NPV), thousand roubles	20632	4709	1814	17890	1739	-1192
Discount payback period, year	1.24	1.37	1.4	1.26	1.41	No
Internal rate of return (IRR), %	38.8	13.3	13.5	32.1	8.50	0.8

4. Conclusion

Based on the study, we can conclude that when structuring and project appraisal of investment projects, it is necessary to correctly identify project risks and objectively assess the likelihood of their occurrence, as well as possible consequences, in order to ensure adequate coverage using adequate tools and techniques. It is very important that the choice of procedures for reviewing projects ensures the maximum neutralization of subjective factors that may accompany credit and investment processes.

From the point of view of the investing decision maker, the ratio of risk and profitability is the basic category that underlies the decision to invest. The developed effective risk identification system for project financing will allow to make an objective decision on the possibility of financing taking into account uncertainty factors.

References

[1] Nevitt P K 1989 *Project Financing* (London: Euromoney)
[2] Wood P R 1995 *Project Finance, Subordinated Debt and State Loans* (London: Sweet & Maxwell)
[3] Fabozzi F J 1995 *Investment management* (Prentice Hall)
[4] Brealley R R, Myers S C and Allen F 2009 *Principles of corporate finance* (Boston: McGrow-Hill/Irwin)
[5] Shenaev V N and Irniyazov B S 1996 *Proektnoe kreditovanie. Zarubezhnyj opyt i vozmozhnosti ego ispol'zovaniya v Rossii* (Moscow: Konsaltbankir Publ.)
[6] Wu L, Gao H and Wei C 2019 VIKOR method for financing risk assessment of rural tourism projects under interval-valued intuitionistic fuzzy environment. *J. Intell. Fuzzy Sys.* **37(2)** pp 2001-08
[7] Yescombe E R 2016 *Principles of project finance* (Moscow: Alpina Publisher)
[8] Kazanskiy A 2015 Riski proektnogo finansirovaniya i metody ih preodoleniya v rossijskih proektah *Problems of modern Economics* **1 (53)** pp 139-43
[9] Yoo J-I, Lee E-B and Choi J-W 2018 Balancing project financing and Mezzanine Project Financing with option value to mitigate sponsor's risks for overseas investment projects *Sustainability* **10** (Switzerland) Issue 5
[10] Natocheeva N N and Belyanchikova T V 2018 Risk management and optimization of bank loans allocation in the project financing program *Journal of Applied Economic Sciences* **13(1)** pp 65-73
[11] Ilin I V et al 2019 Modeling of the projects investment attractiveness implemented under

conditions of project financing *The European Proceedings of Social and Behavioural Sciences* **59** pp 610-19

[12] Nikulina E and Tarasova E 2019 Budget financing of investment projects at the high-technology industry enterprises: Project implementation technology and risk mitigation *International Journal of Recent Technology and Engineering* **8(2)** pp 5932-35

[13] Ma Q and Sun M 2017 Risk assessment of engineering project financing based on PPP model *MATEC Web of Conferences* **100** pp 1-7

IOP Conf. Series: Materials Science and Engineering **911** (2020) 012019 doi:10.1088/1757-899X/911/1/012019

Factors of environmentally oriented development of management technologies in construction

I Nuzhina[1,2,3] and L Shershova[1]

[1] Immanuel Kant Baltic Federal University, 14 Nevskiy St, Kaliningrad, 236016, Russian Federation
[2]Tomsk State University of Architecture and Building, Solyanaya sq., 2, 634003, Tomsk, Russian Federation
[3]Tomsk State University of Control Systems and Padioelectronics, Lenina Prospect, 40, 634050, Tomsk, Russian Federation

E-mail: irinanuzhina@yandex.ru

Abstract. The results of research to advance the study and practice of construction management in the context of greening and transition to a green economy are presented. The basic aim of the study: is exploration of development of management technologies space available to construction managers for the development of a fully interlinked system of indices of the controls as a model of environmentally sensitive and ecosystem-based management processes. Methods of system, logical and comparative analysis are used. The study provides an overview of the institutional environment and identified the factors dictating the necessity to transform technology of project management control in construction in accordance with the environmental regulations of development and the social needs of city-building activity. Has been reasserted the importance of increasing greening needs of the population, social and environmental responsibility of the construction business is shown. The author's concept is proposed and structured system of eco-engineering project as a role model for the development paradigm of technologies for governments in construction was introduced.

1. Introduction

The current stage of development of society is marked by a matter of priority for solution of social problems and their link to improve environmental living conditions. Mainstreaming of the green economic development objectives dictates the need to develop methods and tools for regulating the economic behavior of economic units, including those related to the construction sector. Taking account the significance of town-planning and building activities for upgrading the technological relevance of production facilities, undertaking the task of social affairs and housing problems, and developing the urban environment, it is imperative to ensure good conformity assessment of construction managerial practice to requirements of environmentally sustainable development. To generate further dynamism in promoting green building practice in Russia, management methods and tools, green technical and technological resources, a system of green certification and regulatory technical standard-setting toolkit must be integrate into one system.

The methodology for developing an environmentally consistent management model is based on a multidisciplinary approach, theoretical and practical research by scientists and professionals in various subject areas. Thus, in the field of environmental research the works of scientists Telichenko V. I.,

Benuzh A. A., Slesarev M. V. [1-3], Il'ichev V. A, Kolchunov V. I., Bakaeva N. V., Kobeleva S. A [4,5] should be emphasized. The research papers of Tetior A. N. are devoted to the problems of studying the linkage between greening of the economy and sustainable construction, development of eco-efficient environmental infrastructure [6]. Issue of developing eco-buildings housing market is investigated in the works Krygina A. M. [7,8]. The works of Bobylev S. N., Porfiriev B. N. [9 -11], Pakina A. A., Gorbanyov V. A. [12] are devoted to the study of various aspects of economic modeling in the context of sustainable development, the green economy prospects as a contemporary development paradigm.

Today, in Russia, the challenges of sustainable development of territories come to the forefront of socio-economic system transformations, and the "green" economy is a means of solving them [9,10]. According to the GGEI (Global Green Economy Index) published by the American consulting company Dual Citizen LLC, in 2016, Russia was ranked 51st out of 80 countries (32.59 points) in terms of perception and 74th (38.80 points) in terms of performance [13]. Experts noted low activity in the field of green innovations and investments. In 2018, the rating covered 50 more countries. Russia's position in the overall index did not improve – 105th place with an index of 0.41. the Leaders are Sweden and Switzerland (1st and 2nd place, the index – 0.76) [14].

For the transition to a green economy, it is necessary to implement institutional changes aimed at forming a management model in the context of sustainable development of territories, economic entities and projects. This means that the emphasis in the development of management methods and tools should be shifted from quantitative to qualitative criteria, giving priority to the state of the environment and society [12].

Environmentally oriented management in construction is aimed at integrating methods and tools that ensure the greening of the processes of creating construction products throughout the life cycle. The system should cover on the one hand all the main and supporting business processes, and on the other hand-production and technological processes, investment and architectural design, organizational interaction of participants in the process of implementing construction projects. The problems of forming an environmentally oriented management in construction are not considered in isolation, but in conjunction with the functional management system of the enterprise, the directions of restructuring the construction complex, General trends in organizational and technological development of interaction schemes of participants in construction activities [15-17].

2. Materials and methods

The modern environment of management technologies development in construction is formed under the influence of various factors. Among the defining activities should be the promotion of the "green" economy model as a tool for ensuring the practical implementation of the concept of sustainable development. Today, this is a paradigm of human development recognized by all countries in the twenty-first century. In the context of sustainable development, the world community and Russia have adopted strategic documents defining the vector of development of the economy as a whole, individual sectors and territories [9, 10].

The next important factor is the development of institutions and environmental construction tools. In the context of the studied issues, the concepts of ecological construction and "green" construction will be considered as identical. Significant institutional changes have taken place in the sphere of organization and regulatory support of construction. In the first decade of the XXI century in Russia non profit organizations were created: "Green Building Council in Russia" (2009); Association of manufacturers of quality housing "GreenStroy" (2010); partnership "The Center of Environmental Certification - Green Standards" (2010). The development of construction according to environmental standards was also facilitated by the creation of self-regulating organizations. In 2009, Association "EkoStroy" was established, which started creation of the National Eco-Building Cluster of the XXI century, which unites designers, manufacturers and builders in a single cooperative, whose activities are aimed at promoting the implementation of eco-building projects and the development of environmental certification. In 2015, ANO NIIURS was established to provide services to improve energy efficiency and environmental friendliness of construction projects. In 2016, the Technical committee on standardization of Green Technologies of Life Environment and Green Innovative Products (TC 366) was established.

Due to increased environmental construction institutions activities, regulations (mandatory and voluntary) supporting the green construction technologies promotion in the urban development projects implementation have been devised and implemented. The certification system of information and technical documentation was created: the building materials' certification system "Vitality Leaf "(2001); "Green Standards"(2010); "Ecological Passport"(2010); Rating System of Habitat Sustainability Assessment (2011-2014); NDT reference books (2015); green standards, green technologies of living environment criteria (2019). The latest standards streamline terminology, systematize technologies and set criteria, which allows further developing of the methodology of economic evaluation of green construction technologies – from the object and work to the environment. In 2014, the first standard adapted to Russian conditions was adopted – the GREEN ZOOM system.

Due to the development of the regulatory framework, the total number of certified buildings in Russia has also increased. Until 2018, these were mostly facilities certified using the British BREEAM, American LEED and German DGNB systems. The total number of such facilities for the period from 2010 to 2019 was 130, the total area-3011, 15 thousand sq. m. The most popular system is BREEAM-72%. A significant part of certified objects are office premises (39%), and in Moscow – 80% [18]. The total number of objects in Russia certified under the GREEN ZOOM system amounted to 40 in 2018, and increased to 48 in 2019. It strikes the attention that 71% are residential objects, and 21% are office objects [19, 20].

Another significant factor affecting the development of management technologies in construction is the increasing greening of the population's needs for construction products. This manifests itself in an increased interest not only in ecological housing, but also in a broader aspect – in the urban environment that provides security, comfort, a favorable environmental environment, the preservation of natural landscapes, historical and cultural monuments. The results of research by ANO NIIURS (2019) showed that 57.9% of respondents noted a mediocre and unsatisfactory environmental situation in their place of residence [20]. Moreover, 66.5% of respondents noted the importance of the environmental factor when deciding on the choice of housing. This even exceeds the significance of the factor-the cost per square meter-64.6% (Figure 1). The state of social infrastructure is considered a priority factor for 68% of respondents, the layout of an apartment or house as a factor of comfortable living-68.4% of respondents, 20.2% of respondents noted the importance of environmental certification of finishing materials, household chemicals, and the absence of ozone-depleting components when choosing materials. The results of the survey also revealed insufficient development of social infrastructure. The analysis of the formulated wishes to developers showed that citizens identified the following aspects: reducing the number of storeys as an important factor in improving the urban environment comfort; noise isolation of apartments and houses; creating a barrier-free environment; separate waste collection and awareness of citizens about its advantages; organizing yard spaces without cars; organizing Bicycle infrastructure; creating public spaces for all groups of the population; organizing playgrounds for walking animals and increasing the number and variety of green spaces [20].

IOP Conf. Series: Materials Science and Engineering **911** (2020) 012019 doi:10.1088/1757-899X/911/1/012019

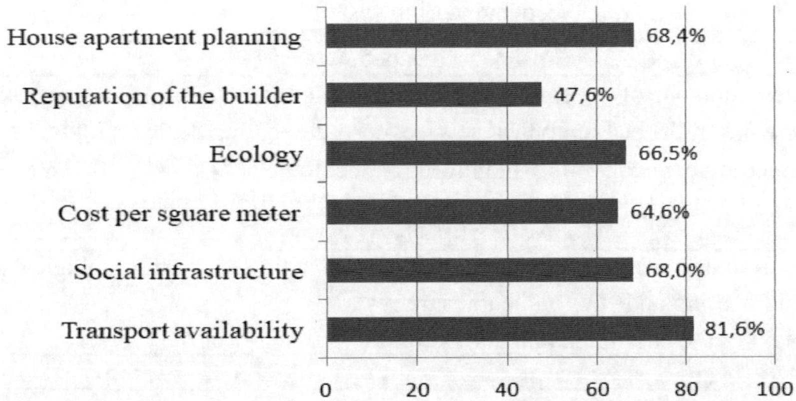

Figure 1. Results of the survey on the importance of factors in choosing housing

Greening the needs for construction products encourages the development of social and environmental responsibility of the construction business as an initiative activity of construction companies, which is expressed in the adoption of additional increased social and environmental requirements for construction projects, works performed and construction objects [21]. It is the responsibility of business that will allow us to find a balance between economic interests, environmental and social priorities for the development of territories in the process of implementing construction projects. A compromise is achievable. On the one hand, it is based on improving the environmental friendliness, energy efficiency and comfort of objects and territories, which ensures an increase in the investment and marketing attractiveness of objects, and the profitability of investment projects. On the other hand, based on the development of regulatory and legislative support for "green" construction (mandatory and voluntary) and financial incentives for environmental construction. Greening needs of the population promotes community involvement in the decision-making processes for construction and development of residential areas. Collaborative design is now becoming another significant factor in the development of organizational aspects of management technologies in construction.

Implementation of the model of environmentally oriented management in construction is impossible without training of professional personnel [22]. The modern paradigm of educational programs is the integration of education technologies for sustainable development into training programs for specialists at all levels, updating the methodology of the ecosystem approach based on the relationship of economic, environmental and social components in the process of solving professional tasks and performing labor functions. Thus, professional training should equip the specialist with knowledge and skills in the field of engineering, which provides for increasing the efficiency of the construction business based on the development of technologies for environmental support of construction at all stages and processes of the life cycle of construction products. Of course, the content of engineering will differ depending on the subject area and will require some specialization of employees.

3. Results and discussion

The identified factors of development of management technologies in construction determine the choice of a methodological basis for modeling the architecture of a system of environmentally oriented methods and tools for regulating construction activities. As such, the project approach should be used, focused on transforming the business processes of construction activities based on the development of ecoengineering of the project (Figure 2).

Environmentally oriented development of management technologies in construction

IOP Conf. Series: Materials Science and Engineering **911** (2020) 012019 doi:10.1088/1757-899X/911/1/012019

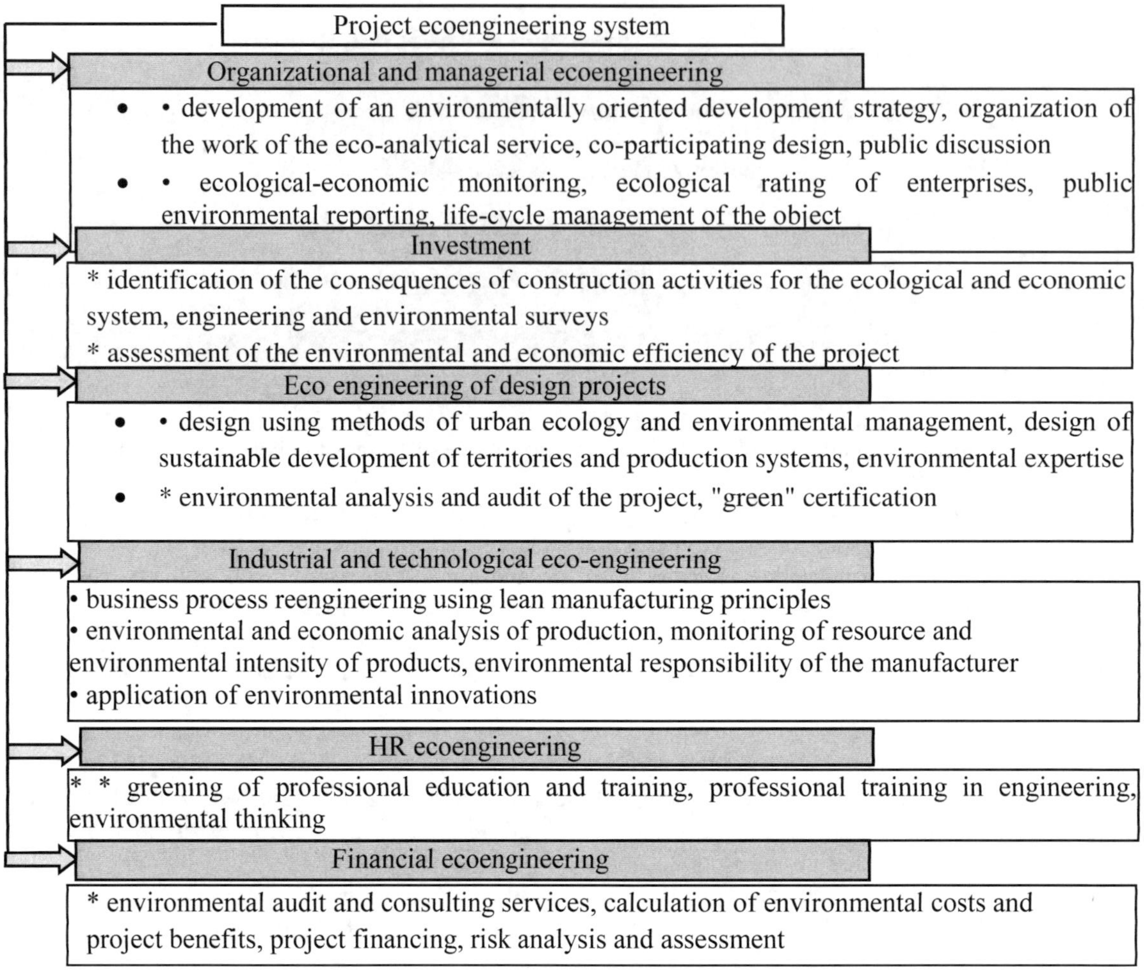

Figure 2. Elements of the project's ecoengineering system

Ecoengineering as a form of improving the environmental efficiency of the construction business specializes in providing services, performing works of organizational and managerial, investment and project, production and technological nature, including personnel and financial support. The shift in focus in management to project-oriented activities reflects the current trend of management development. Production activities of enterprises are implemented in the process of implementing construction projects. Therefore, the functions of ecoengineering are closely intertwined with the functions of project management, planning and organization of production activities of the enterprise and meet the criteria. Of the gpm Global P5 standard for sustainable project management-social, environmental and financial responsibility [23].

4. Conclusion

Thus, the main factors of environmentally sustainable development of management technologies in construction are: activities promoting a green economy model internationally and in Russia; developing institutions and products suitable to the needs of the green building practices; greening the needs of the population requirements for construction services; enhancement of corporate environmental and social responsibility of the construction industry groups; co-participating design of urban areas; greening of vocational education and training of specialists in the field of eco-engineering.

The formation of an up-to-date model for the development of management technologies in construction is build on an ecosystem-based approach that implements the sustainable development principles and project-oriented activities through the organization of an eco-engineering system. The scope of ecoengineering covers all subject areas of the construction project, ensuring that the project outcome responded to the social and environmental priorities of industries and land development.

References

[1] Telichenko V I and Benuzh A A 2016 *Academia. Architecture and construction* (Moscow) **1** pp 118-121
[2] Telichenko V I and Slesarev M V 2018 *Vestnik MGSU* (Moscow) **13 5(116)** pp 558-567
[3] Slesarev M V and Telichenko V I 2018 *IOP Conf. Ser.: Mater. Sci. Eng.* (Novosibirsk) **456** 012126
[4] Il'ichev V A, Kolchunov V I and Bakaeva N V 2018 *Industrial and Civil Engineering* (Moscow) **3** pp 3-11
[5] Il'ichev V A, Kolchunov V I, Bakaeva N V and Kobeleva S A 2016 *Scientific Herald of the Voronezh State University of Architecture and Civil Engineerin* (Voronezh) **4 (44)** pp 169-176
[6] Tetior A N 2008 *Environmental engineering* (Moscow) **2** pp 57-67
[7] Krygina A M and Krygina N M 2018 *Russian journal of housing research* (Moscow) **5 3** pp 353-366
[8] Krygina A M 2016 *Russian journal of housing research* (Moscow) **3 2** pp 113-126
[9] Bobylev S N 2017 *World economy and international relations* (Moscow) **61 3** pp 107-113
[10] Bobylev S N, Goryacheva A A and Nemova V I 2017 *E-Journal Public Administration* (Moscow) **64** pp 34-44
[11] Bobylev S N and Porfiriev B N 2019 *Moscow University Economics Bulletin* (Moscow) **4** pp 3-7
[12] Pakina A A and Gorbanyov V A 2019 *MGIMO Review of International relations* (Moscow) **12 (5)** pp 134-155
[13] Global Green Economy Index^{TM} GGEI 2016 *Measuring National Performance in the Green Economy* Dual Citizen LLC. UPL: http://dualcitizeninc.com/GGEI-2016.pdf.
[14] Global Green Economy Index (GGEI) 2018 *Dual Citizen Introduction* UPL: https://dualcitizeninc.com/global-green-economy-index/index.php#interior section_link.
[15] Kaverzina L 2018 *MATEC Web of Conf.* (ICRE 2018) **212 (**Irkutsk 26-27 April 2018) 08030
[16] Kaverzina L and Cherutova M 2018 *Proc. Int. Sci. Conf. "Far East Con"* (ISCFEC 2018) (Vladivostok 02-04 October 2018)
[17] Buzyrev V, Nuzhina I and Zolotareva M 2018 *MATEC Web of Conf. (ESCI 2018*) **193 (**Ho Chi Minh City 02-05 March 2018) 01007
[18] CRE.ru Market research UPL: https://www.cre.ru/ analytics/78471?Fbclid= iwar2jmigkhg 9e5 rrjeoxfd31mfq4m0tyogip93ilfvo2fdk7k9l62sgagqa4/
[19] Annual report NIIURS 2018 UPL: https://yadi.sk/i/KN8Eq-ch9kti7w.
[20] Annual report NIIURS 2019 UPL: greenzoom.ru/Годовой%20отчет%20НИИУРС%20 2019.pdf.
[21] Nuzhina I, Zolotareva M and Vasileva I 2018 *MATEC Web of Conf. (YSSIP 2017)* **143** (Tomsk 25-27 October 2017) 04010
[22] Shershova L, Kurochkin E and Nuzhina I 2018 *MATEC Web of Conf. (IPICSE 2018)* **251** (Moscow 14-15 November2018) 05015.
[23] GRM P5 *Green Project Management* UPL: https://greenprojectmanagement.org/the-p5-standard/

IOP Conf. Series: Materials Science and Engineering **911** (2020) 012020 doi:10.1088/1757-899X/911/1/012020

The process and project approach as the basis for harmonization of development management system of a construction company

V V Dorofeeva[1], L A Kaverzina[2] and I P Nuzhina[3]
[1]Grand PhD in Economics, Kaliningrad State Technical University, Kaliningrad, Ave. Soviet, 1, Russian Federation
[2]Grand PhD in Economics, Bratsk State University, Bratsk, Makarenko str., 40, Russian Federation
[3]Grand PhD in Economics, Tomsk State University of Architecture and Building, Tomsk, Salt square, 2, Russian Federation

E-mail: d310574@yandex.ru, Dekanfps@mail.ru, irinanuzhina@yandex.ru

Abstract. The article is devoted to development of theoretical positions as well as methodical and practical base for increasing the efficiency of cost management of construction project realization. The modernized organization and economic mechanism of formation and cost management of the construction project realization is proposed. This mechanism is determined as integrated system of planning and control for the cost of construction project realization. The main directions of harmonization of components of company's development due management system due to the process and project approach that combines advantages of value-oriented management and methodology of project management are indicated.

1. Introduction.

There is an active search for adequate and effective mechanisms of companies' development in science of management. Study of the wide range of scientific works [1-12] allows seeing a huge variety of theoretical and methodological points of view about development management of construction company. The deficiency of systematic researches of use of different models and development concepts, interaction and interdependence between various management technologies determines the necessity of forming the mechanism for focused management efforts to ensure quality and inversible systematic changes in company.

Modern researches in field of process management show complexity of management system development. Elements of the process management are integrated with the methodology of strategic management, balanced system of indicators, customer management system and benchmarking creating powerful potential of quality improvements in company [1-4].

An alternative development management approach is the project methodology. Appearing and wide use of the project methodology has necessitated the concentration of efforts at the achievement of determined results of development. Thus, the management of development by realization of projects has turned into in come that help organizations to implement their business plans and achieve strategic orienteers of activity [5]. The popularity of the project approach in practice of development management is conditioned by result orientation (value added, product). Nowadays, development projects are recognized as agents of strategic changes and in fact, they penetrate all elements of the enterprise system [6-8], and methodology of the project management becomes the key to effective implementation of any changes in company [9].

Purpose of the article is to develop methodical recommendations for usage of value oriented management tools in process of construction projects preparation and forming of project portfolio. Sharing usage of the proposed approach and appropriate tools in construction will provide a proper formalization for processes of directing of top management efforts to the growth of capital appreciation (property) of developer-company as a participant of the construction and investment process.

2. Materials and methods.

Appearing of value-oriented management concept was related with necessity of forming a new company executives way of thinking in conditions of growing competition in commodity markets and dynamic development of stock markets in 1980s. The choice of value indicator as main strategic orienteer of company development has helped to approve all goals of company activities by concentration of efforts on key factors of value creation as well as determine the dependence between organization, strategy and enterprise value in the capital market [1]. With time, much more attention has been paid to the problem of value company management by Russian companies. It has been caused by several problems:

1. The specific of Russian management, first of all, based on the fact that Russian managers are concerned on issues of production and the correct direction of funds, rather than the market value of the enterprises they lead. This has led to low capitalization of Russian economics and «phenomenal by world standards fact: balance value of the enterprise significantly exceeds market value» [6].

2. Low investment attractiveness of Russian enterprises. The investment attractiveness is directly depend on the correct assessment of stocks value. So, there is a necessity of active practice spread of Russian enterprises value management aimed to capitalization increase.

Functioning of construction enterprises in conditions of crisis prolongation on the background of turbulent market environment, dynamism and globalization of markets stimulates them to search for resources of leveling negative influence of the environment. One of the modern approaches, which has got a wide spread among international management practice, is the value approach or deviation management [11].

If we consider a construction project as complex, then it can be schematically reflected as shown in the Figure 1.

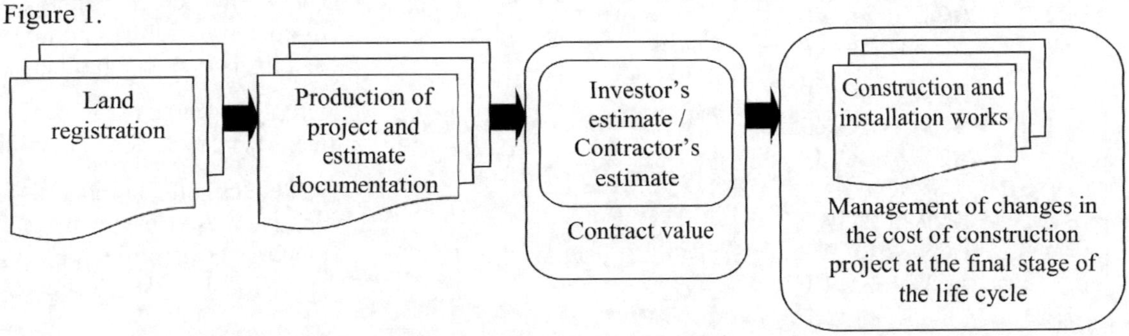

Figure 1. Scheme of the life cycle of a construction project.

After signing the contract, during the implementation of construction project, cost of the construction may be changed due to circumstances not considered in previous calculations or force majeure. In this case, the construction estimate is corrected according to estimated normative documents with use of estimate program complexes, and so, the factual value of particular works and construction process as whole is determined.

The appropriate level of fiscal and value effectiveness of the construction project largely depends on organization of its implementation. It requires optimization of realization duration of individual stages and project as whole, coordination with available financial and material resources and conditions for their use [7].

Project management is the developed methodology, which includes models, methods and program resources that are widely used during development and implementation of projects of different scales and types [10].

3. Results.

Method of mastered volume can be used during the process of modelling on the stage of construction project execution analysis [5; 9; 11]. The relevant report for this method presents the process of project implementation. According to the data of such report, deviations in costs and terms can be revealed. In support of this, several schematic examples of project analysis at a certain point in time are given in [8].

Using the method of mastered volume, problem situations appearing during the implementation of construction projects can be formalized and further events can be predicted. There are some indicators, the content of which need to be understood to efficient use of the mastered volume report in management practice.

Consider 3 main indicators [7-10]:

- BCWS – Budgeted Cost of Work Scheduled;
- ACWP – Actual Cost of Work Performed;
- BCWP – Budgeted Cost of Work Performed.

Accordingly, it is possible to foresee options for possible deviations and determine their causes. The examples of deviations of the main cost indicators are presented in the table 1.

Table 1. Possible causes of fiscal and cost deviations of construction project

No.	Ratio of indicators	Cost characteristic	Time characteristic	Possible causes of deviations
1	BCWS=ACWP BCWS<BCWP ACWP<BCWP	$C^{fact}=C^{plan}$	$T<T^{plan}$	Use of new technologies or mechanisms. Reducing the duration of individual works.
2	BCWS=ACWP BCWS>BCWP ACWP>BCWP	$C^{fact}=C^{plan}$	$T>T^{plan}$	The need to redo individual works. Mistakes in planning of resource providing. Failure to fulfill contracts in time.
3	BCWS<ACWP BCWS<BCWP ACWP=BCWP	$C^{fact}>C^{plan}$	$T<T^{plan}$	Increase resource usage per unit time. Mistakes in time planning.
4	BCWS>ACWP BCWS>BCWP ACWP=BCWP	$C^{fact}<C^{plan}$	$T>T^{plan}$	Reducing the number of employees or mechanisms. Temporary stop of the project or individual works. Resource delays
5	BCWS>ACWP BCWS>BCWP ACWP>BCWP	$C^{fact}<C^{plan}$	$T>T^{plan}$	Replacing individual resources (materials, mechanisms, etc.) with cheaper ones.
6	BCWS>ACWP BCWS<BCWP ACWP<BCWP $U_{ji}>r_{ji}$	$C^{fact}<C^{plan}$	$T<T^{plan}$	Delays in financing. Reducing the lead-time of individual works. Introduction of individual changes.

7	BCWS>ACWP BCWS>BCWP ACWP<BCWP $U_{ji}<r_{ji}$	$C^{fact} < C^{plan}$	$T > T^{plan}$	Tendering for the implementation of individual works (transfer of part of the project to contractors). Temporary cessation of individual works.
8	BCWS<ACWP BCWS<BCWP ACWP<BCWP	$C^{fact} > C^{plan}$	$T < T^{plan}$	Deletion of individual works. Attraction of additional specialists. Introduction of new technologies.
9	BCWS<ACWP BCWS>BCWP ACWP>BCWP $U_{ji}>r_{ji}$	$C^{fact} > C^{plan}$	$T > T^{plan}$	Appearing of extra works in project. Partial change in goals. Prepaid expanse. Purchase of additional resources.
10	BCWS<ACWP BCWS<BCWP ACWP>BCWP $U_{ji}<r_{ji}$	$C^{fact} > C^{plan}$	$T < T^{plan}$	Mistakes of financing management. Prepaid expanse. Conflict of account and production management..

Note: r_{ji} – number of j resource per unit time; U_{ji} – intensity of j resource use, moreover $U_{ji}=1$, it means that intensity is constant; C^{fact} - factual cost; C^{plan} - planned cost; T^{plan} - planned time

Huge problem of the value-oriented management in Russia is the difficulty of use of developed foreign methodologies [13]. Often analytics do not carefully consider and use basic assumptions of these methodologies. At the same time, according to modern accounting rules, intangible assets (trademarks, licenses, company brand and its reputation, etc.) remain unaccounted or carried at cost that does not correspond to market situation. Market capitalization depends on assessments of these indicators' value. This way, if an asset was acquired several years ago, it is possible that its initial value may not correspond to its actual market value; and managers can manipulate certain values, which are reflected in financial statements in order to increase estimated value of such assets.

Cost methodologies of efficiency assessment that have been spread in the territory of the former USSR and within modern systems of resource management (in particular, MRP, ERP, SAP systems) give an opportunity to adequately present and understand the previous experience of realization of business operations [14]. The methodologies of efficiency results assessment , which traditionally have been dominated abroad are getting much more spread in Russia explaining the future and based on assessment of expectations (market,owners, enterprises, competitors, investors, etc.).

During the process of a construction project, management of fiscal and cost deviations of the project where the most important part is the establishment of single information stream from project launching to its introduction into exploitation remains problematic [15]. As many scientists suppose, "management by exception" is one of the most perfect management systems, which based on key indicators of efficiency [16-18]. However, in our opinion, it is necessary to additional take into account the features of modern concepts of "Integrated Project Delivery" and "Building Information Modeling" which determine their specificity of use for operational activity of construction enterprises.

Balanced distribution between key participants and implementers of the project such aspects as reward, responsibility and risks, is set with the system of organization's contractual relationships named "customer-designer-contractor". So, the complex system of planning and resource management of the enterprise must be closely related with variety of aspects, in particular: lineup of contracts in FIDIC construction, informational interaction, digital object models, competencies in area of project management, pricing in construction process, etc. Thus, according to specific activity of a construction enterprise, the process of optimal manufacturing program formation must begin with providing of its production capacity optimization. [20].

Traditionally, during the realization of the project, such important facts as time, resources (productivity, capacity) and cost are considered isolated from each other whereas they are interconnected. So the problem of integrated by time, cost and productivity planning has a fundamental significance for construction project realization. Based on the deviation management methodologies relatively new to Russian science, and practice of value oriented management, which based on planning and control, there is an opportunity to improve the management of construction project realization process that includes cost assessment, cost planning (budgeting) and control of the cost. Such improvement is important in terms of efficiency of construction production since in the existing methodology, cost assessment comes down to budgeting and its management to fixation of factual results and compare them with plan results.

4. Discussion.

As much specialists think, "stage-by-stage movement" of a company in direction to achieve the desired state (in the context of a strategic idea) by realization of projects has significant prospects [4, 5, 7, 9, 15]. At the same time, replacement of the current activity management of an enterprise to project activity, as we suppose, is totally wrongful. Thus, a clear structuring of management decisions area is necessary for companies that are not project oriented. It will help to determine borders of project and current activities and provide coordination of management instruments for rational allocation of resources and time within project portfolios and non-project activity

Apologists of the project concept of development offer an introduction of the project office as the center of strategic development, which in conditions of matrix structure must provide the realization of related tasks [8, 11, 13, 14]. However, such format of project management may be appropriate only within project-oriented 13companies. In companies, which main activity implements in non-project principle, usage of such management system of development is almost impossible [15, 17, 18]. Traditional methods are primary oriented on neutralization of unpredicted costs related to physical objects (construction materials, constructions) whereas most of deviations are caused with human factor, imperfection of the organization, nature, changes, mistakes, delays, which are not assessed adequately in estimates. Because of this, works volume (especially by labor costs) are greatly underrated, so the deviations of value assessment and duration should be classified as optimistic or pessimistic [11].

Conclusion.

Modern business model of integrated of counter action to the fiscal-cost deviations of project within contract construction must rely on new set of procedures, analytical instruments and program products, which together must cover content of processes and works during the whole investment cycle – from initiation, concept development and business planning (including development of alternative investment options), to reaching rated capacity.

So, informational documentation of the project should be divided into 5 groups: schedule, resources, costs, cash flow and reporting. This methodological approach gives an opportunity further to subdivide limited resources into project works thereby determine the duration and cost of works implementation and construction project realization, and also build a flow chart. It significantly extends the range of economic levers and determine the selection of profit as an indicator of efficient activity of the enterprise. According to this, two-dimensional space "scope of work-time" or "scope of work -cost" becomes three-dimensional "scope of work -time-cost".

References

[1] Curtis B and Hefley W 2009 Overview of the people capability maturity model *Software Engineering Institute*

[2] Harker S D P and Easton K D 1993 The change and evolution of requirements as a challenge to the practice of software engineering *USA. Proc. RE'93*

[3] Ivanchina O V 2017 Corporate project management systems *J. Sci. dial.*

[4] Kubasova T I 2015 Current trends in the development of construction project management methodology *Baikal Res. Journ.***4** pp 97-101

[5] Morozenko A A and Krasovsky D V 2016 Management of investment and construction projects based on the matrix of key events (Moscow: *MGSU Bulletin)* **11** pp 105–13

[6] Dorofeeva V V 2001 Main aspects of reproduction of investment and construction process *Bulletin of the Novosibirsk state Academy of construction and architecture* **2** pp 53-57

[7] Bukreev A M and Ryzhkov E N 2010 System approach to the formation of an organizational mechanism for managing the sustainable development of the enterprise's economy *Production organizer* **3** pp 65-7

[8] Gareev I F, Rozhkov V L and Kutdusova A G 2015Main trends in the development of the investment and construction complex *J. Russian business* vol. 16 **19** pp 3149-58

[9] Ryzhakova G M, Prikhodko D O, Predun K M, Lugina T S and Koval T S 2017 Models of target selection of representative indicators of construction companies ' activity: etymology and typology of diagnostic systems *J. Managing the development of complex systems* **32** pp 159-65

[10] Dorofeeva V, Kaverzina L and Koshevoi P 2019 The improvement of company planning at the building enterprise IOP Conf. series: materials science and engineering 667 Investments, Construction, Real Estate: New Technologies and Special-Purpose Development Priorities (Irkutsk Russian Federation 25 April 2019) p 012037

[11] Chernyshev D O 2017 Conceptual bases of organizational and technological reengineering of projects based on the principles of biosphere-joint construction *J. Managing the development of complex systems* **30** pp 205-9

[12] Abramov I V, Algazinov E K and Matveev M G 2014 Process approach to building a project management system *J.Applied informatics* **1 (49)** pp 15-22

[13] Verba V A 2009 Harmonization of process and project approaches to company development management *J. Project management and production development* **3 (31)** p 114

[14] Sharipova O S 2013 Definition of objects of harmonization from the positions of process, situational and project management *J. Project manag. and prod. developm.* **2 (46)** p 89

[15] Shcherbakova N A 2012 Value-oriented management in the enterprise management system *J. Interexpo Geo-Siberia* **3** pp 39-43

[16] Kashina E V and Golovanova L V 2015 Comparative analysis of methods and models of cost approach to resource-intensive enterprise management (Irkutsk: *IrSTU Bulletin)* **5 (100)** p 59

[17] Tevlina S V 2008 Features of project management in construction companies *Scientific and technical Bulletin of the Saint Petersburg state Polytechnic University. Economics* **2 (54)** p 11

[18] Silka D N and Kozulin S V 2017 New tools for managing construction programs and projects in the new structure of the economy *MGSU Bulletin* **11** pp 1214-20

[19] Morozenko A A 2012 Conditions for the stability of an investment and construction project based on the concept of risk management *MGSU Bulletin* **10** pp 260-6

[20] Mazur I I, Shapiro V D, Olderogge N G and Zabrodin A Yu 2009 Investment and construction engineering *J. Economyc* **4** p 74

IOP Conf. Series: Materials Science and Engineering **911** (2020) 012021 doi:10.1088/1757-899X/911/1/012021

Analysis of the requirements of normative and technical documentation for piling equipment

S Plekhanova[1] **and N Vinogradova**[2]

[1] Candidate of Technical Sciences, Associate Professor of the Department of Materials Science and Commodity Expertise of the Russian State University named after A.N. Kosygina, Moscow, Russian Federation
[2] Senior Lecturer, Department of Integrated Security in Construction, Moscow State University of Civil Engineering, Moscow, Russian Federation

E-mail: lisa-xumuk@yandex.ru

Abstract. The article is about piling equipment. Piling equipment is a type of construction equipment that is used to drive piles, drill holes, build a pile foundation and perform other construction work. This concept includes several types of equipment. It is classified by appointment, by power, on attachments and basic equipment, on performance. Pile driving technology is used in modern construction to carry out a wide range of construction works. Depending on the destination, it is worth choosing a piling machine of a certain power and performance.

1. Introduction

Pile driving equipment became widespread in the second half of the 20th century and, with its appearance, made significant changes in construction. After all, before its appearance, the clogging of supports to a depth of 10-12 meters for the construction of foundations in places with loose soil was almost impossible [1].

The supports clogged into the ground by a piling machine significantly compact the soil around and inside the pile field. Thanks to this, large civil and industrial facilities can be erected on soft soils.

Figure 1. Type of piling equipment

The pile construction method is quite common, especially in the north on frozen soils, because during operation, the upper layers warm up, which can cause soil movement. When using piles in construction, part of the load is transferred to more dense layers of soil located in depth. Piles can be used to create trenches and pits to strengthen walls and prevent soil erosion. For driving piles to the desired depth, a piling installation is used. Piles prevent dangerous deformation of soils, leading to uneven subsidence of objects and the appearance of cracks.

Due to its advantages, piles have effectively proved themselves during construction on problematic soils with low strength, weakness, heaving and a large depth of freezing (for example: loam, clay, sand, sandy loam, etc.). The strength of the foundation on piles does not depend on uneven terrain and the level of groundwater. Pile foundations can be built almost everywhere. The exception is the terrain with rocky soil, because stones (or any other large impurities) will prevent the installation of piles. Piles are used not only in works with the foundation of buildings, but also for strengthening the supporting soil layer, in road construction, construction of bridges, etc.

2. Piling equipment classification

Basically, piling equipment installs reinforced concrete and metal piles [2].

There are several types of piling installations. Conventionally, there are 2 types of classification - according to the type of movement and the principle of work.

According to the type of movement, rail, tracked and wheel are distinguished. The latter are distinguished by maneuverability and a sufficiently high speed of movement. The first two types are more powerful than wheeled piling installations, however, they require additional equipment for moving.

Piling installation includes the following units: power plant, hammer and pile driver. The latter is responsible for installing the pile and keeping it in the right position, and the hammer with the power plant ensures that the pile is immersed to the desired depth.

According to the principle of operation, piling installations are: mechanical (M), steam air (P), hydraulic (G), diesel (D).

A mechanical hammer consists of a heavy metal drummer ("baba"), moving along the guides. The principle of operation of this mechanism is as follows: cable-block mechanism, raises the drummer 4 - 5 m, from where it falls under its own weight pile head.

In modern construction, mechanical hammers are rarely used because of:

- low efficiency;
- low frequency of blows;
- inefficiencies when driving piles at an angle.

Steam hammers are driven by steam or compressed air, directly affecting the shock part. Steam (or compressed air) flows through pipelines into the working cylinder of the hammer and, acting on the piston, carries out lifting and acceleration falling parts of the hammer. The energy and frequency of impacts are determined by the course and mass of the body and not are regulated. Driving down inclined piles is not very effective [3].

Team hammers have their drawbacks: low efficiency, bulky steam generating or compressor equipment.

The main advantage of this type of hammer is the environmental cleanliness of steam and compressed air.

The next type of equipment is hydraulic piling installations. They possess higher efficiency in comparison with other types of equipment. Hammer impact power amplified by pressure from the hydraulic system. It makes it easy to adjust the power impact depending on the ground. This type of equipment can be used in conditions dense buildings in the vicinity of other buildings. On par with steam hammers, hydraulic piling installations allow you to install piles not only strictly vertically, but also at an angle. Maintenance of hydraulic piling units more complex, and their cost is higher than previous types of equipment. To work with this view technicians require a highly qualified operator.

Diesel piling equipment operates as follows. Cylinder with moving up and down a baba is installed on the top of the pile. When a baba moves downwards, air is compressed, which, when compressed, heats up and ignites fuel entering the chamber. Gas forces push the baba up, and as soon as the cylinder reaches its upper limit, it begins to fall again - the cycle repeats. Further work runs automatically until the fuel supply stops.

Figure 2. Diesel piling equipment

The advantages that ensured the greatest distribution of this equipment:
- energy autonomy;
- mobility;
- simple and reliable design;
- high performance.

Disadvantages:
- soil vibration;
- difficult working conditions due to dust.

The choice of the type of piling installation is determined by the type of work, soil, climatic conditions. As well as the amount of planned work.

For large objects or objects where you need to exert great strength, suitable equipment on caterpillar or rail tracks.

Rail piling installation is the most powerful and able to provide significant impact load. However, it is also the most difficult to install and requires additional equipment to move. Therefore, crawler-mounted piling installations are used for a large amount of pile work and for driving pylons longer than 12 m.

Pile-driving installations on caterpillar tracks are more powerful than wheeled vehicles, but their speed of movement is slow, so an automobile platform will be required for fast movement over long distances. Crawler equipment helps out in areas with difficult terrain, as well as when working in confined spaces [3].

Wheel-mounted piling installations have great mobility and maneuverability. They are used when working on dispersed objects with a small volume, for example, in the construction of panel-type residential buildings or low-rise construction. They are equipped with both hydraulic and diesel hammers. The base is usually KAMAZ or the Urals. Due to the large size of the base, such a technique is difficult to apply in conditions of dense development.

IOP Conf. Series: Materials Science and Engineering **911** (2020) 012021 doi:10.1088/1757-899X/911/1/012021

Special attention should be paid to small-sized self-propelled piling units, which can work with both screw and driven piles.

Due to its small dimensions, it is convenient to transport equipment, and it can work even in cramped conditions. Due to the installed hydraulic equipment, a large working capacity is achieved without an increase in size. Self-propelled piling installations have good stability and work effectively on hard and medium soils on rough terrain.

Figure 3. Small-sized self-propelled piling units

3. Normative and technical documentation, which regulates the requirements for piling equipment

The main regulatory document governing the requirements for piling equipment is the “Technical Regulation on the Safety of Machines and Equipment” [4].

Technical regulations of the customs union are valid in five states [5].

The technical regulation contains: definitions of applicable terms; rules for circulation on the market or commissioning of technical regulation facilities; safety requirements; conformity assessment procedures for types of vehicles (chassis), single vehicles, vehicles in operation, types of vehicle components; requirements for labeling products with a single sign of product circulation on the market of the Member States of the Customs Union; protective clause; final provisions on the application of documents certifying compliance received before the entry into force of the technical regulation.

This document sets out the safety requirements for machinery and equipment during the design, manufacture, installation, commissioning, operation, storage, transportation, sale and disposal. Piling equipment is subject to mandatory confirmation of compliance with the requirements of the Technical Regulations of the Customs Union in the form of a declaration of conformity.

Safety requirements for piling equipment are established in GOST 31551-2012 "Piling equipment. General safety requirements" [6].

It reflects:

- general safety requirements for the operation of pile hammers;
- requirements for the main structural elements;
- monitoring the implementation of safety measures at the construction site;
- operating instructions;
- equipment transportation requirements.

The fulfillment of safety requirements for piling equipment of serial production is monitored during the acceptance and periodic tests in the amount determined by the regulatory documentation for equipment of a particular type.

Figure 4. Piling equipment at work

Noise characteristics of piling equipment are measured with a sound level meter. Measurement methods are made in accordance with GOST 12.2.026 "Woodworking equipment. Construction Safety Requirements".

4. Manufacturer instructions

With each piling installation, the manufacturer must provide the following instructions: instruction manual, list of spare parts, instructions for maintenance and repair, if necessary, transportation and installation instructions.

On hammers of specific models, indicators such as: mass of the shock, frequency of blows, estimated impact energy, structural weight of the hammer (dry, without headgear) overall dimensions, fuel consumption, working fluid, air or steam.

The design of the hammer must comply with GOST 31550-2012 "Pile driving hammers. General specifications". It provides:

- hammer operation in an inclined state up to 20 ° from a vertical position;
- reliable launch of the hammer in no more than three attempts when pile is upset per impact up to 150 mm;
- hammer control and daily maintenance by one person.

Each hammer must be affixed with a label in accordance with GOST 12969 "Plates for machines and instruments. Technical requirements".

The ability to use equipment for certain climatic parameters and the placement category of the hammer is established in accordance with GOST 15150 "Machines, devices and other technical products. Versions for different climatic regions. Categories, operating conditions, storage and transportation regarding the impact of climatic environmental factors" as agreed with the customer.

IOP Conf. Series: Materials Science and Engineering **911** (2020) 012021 doi:10.1088/1757-899X/911/1/012021

Figure 5. Installation of piling equipment

On the Russian market there are many samples of piling equipment that differs from each other in its type and operating parameters. When choosing a certain type of piling machine, it is very important to take into account the class of work to be done and the condition of the soil (climatic conditions) at the venue. The volumes of construction for which the selected unit should be designed, as well as the possibility of its repair. Only taking into account all these factors, construction work can be carried out at a high level and completed strictly according to plan. The reliability and durability of the foundation under construction depend on the correct choice of these mechanisms.

References

[1] Frei H and Herrmann A 2018 Reference builder. Construction machinery, constructions and technologies *Technosphere* pp 248-259.
[2] Kozlov D V and Novoseltsev Yu I 2005 *Piling equipment* Tutorial (St. Petersburg) pp 2-23.
[3] Kosolapov V G 1974 Pile work (2nd ed., revised and additional) *Vocational education. Construction works* (Moscow) pp 148-189.
[4] *TR TS 010/2011 Technical regulation of the Customs Union "On the safety of machinery and equipment"* 2011 pp 14-25.
[5] Vinogradova N A 2019 *Text. tech.* **3 (381)** pp 61-64.
[6] *GOST 31551-2012 Piling equipment. General safety requirements* 2012 pp 2-10.

AUTHOR INDEX

Abzayev, Yu A 41
Anikanova, L A 25
Azin, A V 1
Bondarev, B 72
Bryantseva, I V 102
Bunkov, V E 41, 47
Burnashov, E M 19
Chernousov, N 72
Chubarenko, B V 19
Dmitrieva, M A 1, 30
Domnin, D A 19
Dorofeeva, V V 90, 114
Esiukova, E E 19
Evsyakov, A S 7
Fedosov, S V 7
Galyautdinov, D R 67
Galyautdinov, Z R 67
Kasharin, D V 85
Kasharina, T P 85
Katnov, V E 35
Kaverzina, L A 90, 114
Klopotov, A A 41, 47
Konovalova, V S 7
Kudyakov, A I 13, 25
Kumpyak, O G 67
Kurgan, A 41
Kurochkin, E Yu 80
Kurochkina, N Yu 80
Leitsin, V N 1, 19, 30
Lotov, V A 25
Matskevich, N A 41
Mirzaev, O O 35
Nikiforova, V A 90
Nuzhina, I P 114
Nuzhina, I 108
Pervushina, D M 25
Plekhanova, S 120
Plyaskin, A S 41, 47
Polyakov, D A 85
Ponomarev, C V 1
Potekayev, A I 41
Pustovgarov, V I 96
Rikkonen, S V 1
Rumyantseva, V Eu 7
Ryabtseva, N E 13
Savintceva, M E 47
Sharanova, A V 30
Shershova, L 108
Shleenko, A V 61
Sivak, E E 61
Stepin, S N 35
Steshenko, A B 13
Sturova, V 72
Tovpinets, A O 19
Ustinov, A M 41, 47
Vinogradova, N 120
Vlasov, V A 41, 47
Volkova, O V 25
Volkova, S N 61
Voronina, N V 102
Xolmurodov, T A 35
Zborovskaya, M I 53
Zimnyukov, V A 53

Institute of Physics
Dirac House, Temple Back
Bristol BS1 6BE UK

ISSN: 1757-8981
ISBN 978-1-7138-3166-2

CURRAN ASSOCIATES INC.
proceedings.com